ALSO BY JAIME S. CARVALHO

Life: its physic and dynamics

From stars to societies: a process in time

THE FORCE BEHIND EXISTENCE

THE FORCE BEHIND EXISTENCE

— TOWARD A NEW BIOLOGY —

JAIME S. CARVALHO

INFINITY
PUBLISHING

ISBN 978-0-7414-8228-0 Paperback
ISBN 978-0-7414-8229-7 Hardcover
ISBN 978-0-7414-8230-3 eBook
Library of Congress Control Number: 2012923672

Printed in the United States of America

Published February 2013

INFINITY PUBLISHING
1094 New DeHaven Street, Suite 100
West Conshohocken, PA 19428-2713
Toll-free (877) BUY BOOK
Local Phone (610) 941-9999
Fax (610) 941-9959
Info@buybooksontheweb.com
www.buybooksontheweb.com

To my wife,

with the recognition that so little is devoted here
for so much always received.

Whatever is truly universal is unique,
and whatever pretends universality cannot admit challenge.

—L. Whyte

CONTENTS

ACKNOWLEDGMENTS

Many people have contributed in various ways to the writing of this book, and I am very grateful to all of them. In particular, I am deeply indebted to Dr. James Watson for introducing me to unitary field theory. I also express my thanks to all the authors, the vast majority not referenced, that have deepened my knowledge on so many issues discussed here. In addition, I would like to thank the Providence VA Medical Center for allowing me to use its library facilities over the years, and my thanks are extended to the librarian staff, particularly to Ms. Cheryl Banick and to the late Mr. Timothy Gormally for their courtesy and essential help. I also would like to express my appreciation to my nephew Vitorino for our long philosophical discussions, and to my dear niece Maria Antónia for her continued interest and expert help with the computer. Thanks also go to my son, Alex, and my daughter-in-law, Laura, for their unabated support, and to my grandchildren Michael, William, and Katherine for giving me inner strength. Finally, I want to express a special debt of gratitude to my lifetime partner and colleague, Angelina, to whom I owe much of whatever I have achieved in life and to whom this work is dedicated.

INTRODUCTION

Since primitive times, some relation between us and the surrounding universe was suspected to exist, but its nature always escaped human understanding. The most influential Greek philosophers considered the external world to be in constant, chaotic change, a thought that spread over the centuries until the advent of science. This assumed fact was difficult to reconcile with the presumed permanence and harmony of our interior world of ideas. It turned out, however, that behind the apparent chaos of the particles of the real world there is an underlying order of a special type, in which we partake. The scientific explanation of this kind of order was achieved only in recent times, from insights taken not from physics but from biology. The theory—the most general ever conceived—is still in its early stages, but it can already be a great help in the understanding of the universe and of ourselves.

The order embedded in the universe always impressed astronomers and physicists alike, but the simplicity of the celestial systems and the high speed of the physical reactions did not favor its comprehension. The source of that order was easier to recognize in the very complex organisms, where the low energy of their chemical reactions allowed more time for observations to be performed.

Another block for physicists in the road to understanding the natural order was the static concepts they inherited—among those, that of *matter*. Whereas energy implies movement, matter implies permanence and stability; energy therefore was conceived as disembodied from matter. Under the concepts of the new, biologically derived physics, disembodied energy does not exist; energy is incorporated in matter, in the form of any asymmetry—of position, orientation, rotation, polarization, spin, etc.—making up what is called *structure*. The term *energy* was replaced by that of structural asymmetry, and the new term *structure* conveys now the idea of change.

In present-day physics, one of the conspicuous concepts of permanence is that of conservation of mass-energy, which forms the basis of thermodynamics and the foundation of present biology. In reality, however, mass-energy is not conserved: there is always some energy loss in every transformation of structure. Mass-energy appears to be conserved in the chemical reactions of biology simply because they involve only the peripheral electrons of atoms, whose mass-energy content is miniscule compared to the overwhelming bulk of mass-energy contained in their atomic nuclei. The lost energy is of the order of parts per billion only and thus escaped Lavoisier's measurements. But for the formulation of an exact theory of science, particularly for one that expresses the law of nature, this energy loss cannot be neglected.

The concept of conservation of mass-energy expresses the assumption by physical science of the equality of the cause and effect relation: in every natural phenomenon, under isolated conditions, cause always equals effect. As a consequence of this assumed equality, we have a reversible physics. But scientific knowledge accumulated over the last two hundred years strongly suggests that the physical relation of cause and effect is not equal. A one-way or irreversible physics is in operation in our universe.

Present reversible physics works for simple physical systems where cause and effect are approximately equal. Taking the case of the solar system as an example, it does not matter to the laws of motion whether the planets rotate to the right or to the left. But the same does not occur with the extremely more complex biological systems. If we take a good look at a fertilized ovum, we see it growing from inside to outside, and the same occurs to the newborn baby. Development occurs only in one direction. This common biological fact can only be accounted for by an irreversible physics.

Moreover, biological development is continuous, although several phases can be discerned in the course of time, such as growth, maturity, and decay, ultimately followed by death. And the same occurs with physical systems, such as stars and galaxies. But continuity of change implies the existence of an intrinsic order; otherwise continuity could not be sustained. We take this order for granted when making plans today for vacationing in the summer of next year, because we know for sure that the latter will arrive in its due time. An orderly continuity of this type is called a *process*. There is thus an invisible process working in nature and in any one of us. This still largely unknown process is the subject of this work.

On logical grounds alone, we can understand that there should be one process only, because in the case of more than one, they would inevitably clash at some point in the course of time, and this would threaten continuity. But a single process may have two alternating components making up a cycle, which repeats itself indefinitely or until some external influence stops it. In biology, these components are identified with catabolism and anabolism, which are extensions of a similar but simpler process operating in nature. Anabolism is associated with the buildup of structure and catabolism with destruction of the fuel molecules imported from food supplied by our environment.

The structural basis of life rests upon a water-protein three-dimensional structure containing a multiplicity of electrical dipoles—it is a polarized, and polarizable, structure. And the sum total of all polarized structures in the body constitutes the biological field. This polarization field originates in, and is sustained by, the constantly changing structure. Structure and field are thus intimately related and in effect indistinguishable. The new organic physics unifies structure and field naturally, a feat never achieved by any other physical theory.

Since proteins are molecular structures made up of atoms, and atoms are more or less stable aggregations of physical particles, the biological field is an extension of the physical field. More specifically, the biological field is the physical field operating at a higher level of complexity. Because there is only one field and one form of process, the theory is called *unitary field theory*.

Scientifically described by Lancelot Whyte, a Scottish theoretical physicist, in the first quarter of the last century, and further developed by Leo Baranski, an American physicist and psychologist, unitary theory is the only general theory available and as such is capable of describing all other existent theories, which are treated as special cases.

Fundamentally, the theory is based on structure itself, conceived as a system of asymmetrical relations. It uses a concept of time—a local time—different from the relativistic one, which does not apply to human beings. Contrary to the ever more complex current physical theories, the new physics is based upon a simpler foundation concept. In addition, it considers the field and the environment as active agents of the living process.

This new concept of reality brings nature, environment, and organism into a single whole. And for the first time in human history, our psychological processes are identified as expressions or manifestations of the universal field process. The dualism of body and mind is scientifically explained.

The unitary theory brings a new approach to the facts about organisms, requiring some readjustments to be made in our concepts about biological structure and its modification in disease processes. If it is found to be correct, the theory opens up new avenues for research, which should be guided by unitary thinking.

Furthermore, our direct relation to nature and the unitary concept of order impart a significant meaning to us. They contain implicit esthetic values that, when properly understood, can become the seeds for unification of the human society. This is of extreme importance at a time when humanity faces unprecedented challenges, in significance and extent, that only cooperation among all of us can overcome.

THE FORCE BEHIND EXISTENCE

1

THE UNIVERSAL PROCESS

There is general agreement that the basic component of our universe is energy. What is less certain is the nature of energy itself. And in this uncertainty lies the mystery of our existence.

We are all well acquainted with heat, which is a form of energy, but the energy making up the universe has qualities different from those of heat.

Nobody knows exactly what universal energy really is. This invisible, immaterial, and ultimately unimaginable quantity eludes the human intellect. So, for physics, the problem has always been the relation of energy to matter, which is considered to be a stable, unchangeable entity.

Classical physics describes energy as matter in motion, the capacity for movement being ascribed to energy; and more recently, special relativity has identified energy with mass, which is a property of matter.

According to the physical theory to be described, energy outside matter does not exist. Energy is somehow embedded in matter, which must be transformed in some way for energy to become manifested. This view of things makes the classical concept of matter as unchangeable substance obsolete.

The concept of structure

The rigid concept of matter, with its connotation of "no change," has been replaced by the more fundamental and flexible concept of *structure.* Matter implies permanence, and structure implies change. But the change in structure is not that of movement.

In general terms, structure refers to the arrangement of elements in a substance, body, or system, but this arrangement is not arbitrary. For the moment, we may think of structure as a system of relations obeying certain logical properties.

The relations in structure are not substance-bound, and they may be of diverse nature. Here, we are most interested in the spatial and temporal relations, which are the ones most important to physics. Among the spatial relations are those of size, shape, orientation, and relative position.

But the question remains, in what way is energy embedded in structure? And what changes must occur for energy to be released from structure? These are obviously very deep questions, ones that go down to the very heart of physics.

Perhaps even more challenging than the question of release is that of storage of energy in structure. We all know that there is a lot of energy in a chocolate bar, yet the substance chocolate is stable for years. In fact, energy can be stored in structure for millions of years without apparent loss. This is the case of the fossil fuels, which are also deposits of energy.

It may come as a surprise to the reader to know that insights into these questions came from the study of biological structure rather than physical structure. This may appear to be a contradiction of sorts, but events are more delayed, and therefore more readily observable, in the realm of the low-energy biological structure than in the realm of the high-energy and fast-changing physical structure.

It took more than 2,500 years of thought and experiment for us to reach a scientific understanding of the energy problem and its relation to structure. It is a fascinating topic, but to initiate its discussion, we must bring to the fore still another cardinal concept—that of *form* and its development.

The concept of form and formative process

We need words to express our thoughts, and as thoughts develop and change over time, so does the meaning of the words expressing them. This is clearly illustrated by the word *form.*

Form was first conceived in Greek philosophy as the essence of things. But forms meant different things to different philosophers. To Pythagoras (c. 580–500 BC) they were numbers, and to Plato (c. 428–348 BC) they were ideas. Spatial forms were first identified by Aristotle (384–322 BC). He was able to distinguish form from matter and even recognized a formative principle by which the potentialities of things were realized. He maintained that in a formative process, the "whole" appears first and somehow determines the properties of the "parts."

About AD 1600, with Galileo (1564–1642), Kepler (1571–1630), Descartes (1596–1650), and Newton (1643–1727), the earlier interest in spatial forms was displaced by the mathematical analysis of quantities, and the concept of form was degraded from essence to just consequence of the behavior of the material parts of things. Mechanical interaction between static parts constituted the formative process.

After 1750, there were great scientific achievements in all branches of science, particularly in biology, chemistry, thermodynamics, and theoretical physics. As a result of the insights obtained from all these disciplines, there was a shift in philosophical and scientific thought from static to process concepts. A formative process began to be understood as the development of form, from an unstable beginning to a stable end—process was one-way. Instability was thought to be caused by differences of various types, which acted as forces, and stability was found to be associated with spatial symmetry. The role of geometry in physics was beginning to be appreciated.

By the end of the eighteenth century, the study of process as a pattern of changing relations was well under way. And a few decades later, Bertrand Russell (1872–1970) emphasized the fact that it is the space-time relations of structure—the spatial forms—that our perceptions infer. Since then the concept of form became intimately related to that of process.

Because of its continuity, process cannot be seen, and therefore the human mind cannot conceive of it. But the existence of a process in nature cannot be denied. Just look at a tree. You can see it grow over time and change its form. In the winter, it does not appear to be growing, but something is changing inside a tree, which will be manifested in the following spring. Then growth becomes evident, and new forms with different qualities appear in the shape of leaves, flowers, and fruits. There is continuous change in a tree because it is a living system. Although we cannot see process itself, we can see the specific forms in a given phase of process.

But how is the living process generated? What occult "force" is at play? And how is it sustained and directed? The answer to these questions was still not at hand. The curtain of ignorance was beginning to be raised, but not enough light was illuminating the conception of process.

Some kind of link was long ago suspected to exist between nature and our minds. But when concepts of permanence, such as those of economic survival and reproduction, dominated human thought, the idea of process as incessant change could not be grasped, and the above link always escaped rational explanation. Attempts to unravel its secrets drove the philosophers crazy for centuries. But now that process had been discovered and accepted, at least by some prominent scientists, the operation of nature's formative process was still not readily apparent. It was beyond the scope of the scientific method of the time, which could only cover the permanent and quantitative aspects of phenomena.

According to the principle of causality of classical physics, an effect must have a cause. And the most fundamental characteristic of any science is the type of cause-effect relation it assumes. Up until the end of the nineteenth century, it was believed that, in any isolated system, cause always equals effect. On this view, the causal continuity of an isolable (on the way to isolation) process was, in the saying of Whyte, "a perpetual sequence of equal causes and effects constituting an aspect of unchanging permanence within the changing appearance." Equality of cause-effect relationship was at the basis of the principle of conservation of energy, one of the pillars of the science of the time and of present science as well.

Then, in a stroke of genius, Pierre Curie (1859–1906) changed all that. On pursuing the thoughts of previous workers, he reached the conclusion that effects can be more symmetrical than their causes (but not the other way round), without violating the principle of causality. This radical shift of ideas, from symmetrical to asymmetrical cause-effect relations, was an essential step toward the scientific description of the formative process operating in nature.

The formative process of the unitary field

Energy possesses a unique characteristic: it cannot be produced de novo or destroyed. It can only be transformed into other forms of energy. By continually extrapolating this thought, we reach the conclusion that there must exist an initial source of energy from where all other energy forms derive. In physics, this ultimate energy level is called the *field*.

The field is not accessible to direct scrutiny, either by the naked eye or by any present or conceivable means of observation. The field, however, is what fills the whole of the universe. Stating it more clearly, the field is the universe.

For reasons not to be discussed here, the field theories accepted by present physics are all of great complexity, and there is no fundamental change of concepts among them. More importantly, none of these theories is applicable to biological systems. They are thus not general theories. For a theory to be general, it must be of extreme simplicity, it must be able to accurately predict the behavior of the real field in all realms, and it must be capable of explaining all other partial theories. A field theory fitting these requirements is the one scientifically described by the Scottish theoretical physicist Lancelot Law Whyte, in the middle third of the last century, and later expanded by Leo John Baranski, an American physicist and psychologist. Whyte named it *unitary field theory.*

Being a theory of process, its foundation concepts are very different from those of the static field theories of present physics. They represent the synthesis of all scientific knowledge in mathematics, physics, and biology thus far obtained. For these reasons, unitary theory is not just one more field theory. It is the most general, profound, and challenging physical doctrine ever conceived. When applied to present science, it makes us realize how little we know fundamentally.

As its name indicates, unitary theory considers the universe to be just one field, which is made up of infinitesimally small particles, far smaller than electrons. Besides having intrinsic motions, these sub-electronic particles also have intrinsic formative properties of their own: they provide the potentialities for structure to develop.

The most basic of these particles are considered to be *free-energy particles*. But they are not point particles or strings; instead they are concrete three-dimensional structures. The unitary field is thus a quantum field and not a continuous entity.

The free-energy entities of the physical field are polarized structures, in the sense that they display asymmetry. The latter is seen as a distortion from a particular type of latent symmetry. In this polarized field, free-energy corresponds to *structural asymmetry.*

For convenience, structural asymmetry can be regarded as a vector quantity and, as such, capable of being represented by an oriented arrow located in the system. This arrow defines the deviation or distortion of the system from a specific type of symmetry.

The free-energy asymmetrical structures of the field have an intrinsic tendency to symmetry, which the field is constantly negating. But this tendency to symmetry can be realized if the asymmetrical structures become momentarily isolated from the field. Under these circumstances, the symmetry tendency leads some of these structures to aggregate and form new, more complex but also more symmetrical patterns of structure. The tendency to symmetry, however, does not go on for too long, since the field as a whole tends to bring the newly formed structures to the asymmetry level (free-energy level) of the field.

The unitary process has, therefore, two aspects: tendency of structure to symmetry, and tendency of field to uniformity of asymmetry. The first tendency is contained in the *unitary principle* of Whyte, which states that "asymmetry tends to disappear, and this tendency is realized in isolable processes"; the second tendency, the field tendency to bring all formed structure back to the asymmetry level of the field as a whole, is known as the *normalizing process.*

These two tendencies or forces always act together and cannot be separated from one another. They are in fluctuating balance, and their actions complete a circle between them. The first tendency, the downswing of the circle, is associated with transformation of internal structure, leading to a new, more symmetrical form. It was by a series of structural transformations of free-energy structures that increasingly more complex forms—quantum field structures, protons, neutrons, atomic nuclei, atoms, and molecules—were generated. Change in unitary theory is therefore not movement but *development of form,* and process is no more than form within change. Only when its specific form is recognized can a process be identified.

The tendency to symmetry is time-invariant, and it is this invariance that assures the continuity of process. There is also a time delay on going from earlier states of structural asymmetry to later states of less asymmetry. This temporal succession defines a *local time,* which is different from the time concept of present physics.

Included in the pattern of any living structure there is a latent memory, a record of the process by which it was formed. This historical record is not lost when the structure is further transformed. Rather, the information of the past continues and molds the information of the present.

Without further analysis into the transformation of structure carried out in isolable processes, we can sense that there must be a meaningful order underlying change. Disorder tends to decrease, or an increasing order is reached as symmetry is approached. The unitary formative process is thus an *ordering process.*

The second tendency, the upswing of the circle, supplies structural asymmetry to the newly formed structure to bring it up and sustain it at the asymmetry level of the field. The more symmetrical structure has a lesser amount of free-energy than its asymmetrical counterpart and will require less free-energy to be brought up to the asymmetry level of the field. Thus, the formation of progressively more symmetrical and ordered structure increasingly *facilitates* the normalizing process. This hidden economy of nature favors continuous one-way development.

From the two tendencies, it is the second that is unique to unitary theory. It is not contained in the laws of thermodynamics or in any other physical law. Its discovery came from a deep insight into biological metabolism. It turns out that what we call anabolism and catabolism in biology are just highly differentiated expressions of the two unitary tendencies. The first tendency is manifested in anabolism and the second tendency in catabolism.

The formative process generates structure with new qualities and is therefore creative. Since it is universal, it manifests itself in all physical, biological, and sociological systems in the universe. What we call evolution, life, and consciousness are just different aspects, or expressions, of this universal process.

This creative process has been working undercover on the surface of our planet for a long time, building up and organizing ever more complex structure along the one-way evolutionary development. And it was just waiting to be unveiled by unitary science.

Implications of a formative process in nature

It was an astonishing feat of the human mind to conceive and scientifically describe the process whereby nature operates. In terms of power and importance, this accomplishment ranks higher than the discovery of the atomic bomb or the genetic code. The logical explanation for such a remarkable achievement resides in the natural process itself: it directly connects the basic field to our thinking organ. Certainly, there can be no limits to human understanding.

One of the consequences of a single formative process in nature is that every form in the universe, simple or complex, is related to every other form. We are related to the stars, to flowers, and to bees because all those forms ultimately derived from the unitary field along an invisible chain of structural transformations. The inorganic and organic worlds are intimately related. For instance, all of our food ultimately comes from the synthesis of carbon dioxide and water achieved in the process of photosynthesis that takes place in green plants. And the green plants got their photosynthetic ability from primitive bacteria. In a certain sense, we are no more than bacteria, developed and transformed.

Forms can exist in two states: *static forms* and *process forms*. Static forms are those that have reached their maximum symmetry and are now stable. This does not necessarily mean that they have reached complete, or final, symmetry; it only means that their internal structure is no longer changing, but they can still extend their form by the process of duplication. This is the case of our genes and proteins. The concept of structure is properly applied to this form state. Conversely, process forms are those that are still changing their internal structure. This is the case of our psychological memories.

Then there are all the human-constructed forms—buildings, cars, books, toys, and the like—which are permanent things since their very beginning, for they never were in process. They are not living things, and so they are outside of the process of nature.

The widespread existence of man-made permanent forms tends to obscure the concept of process. We tend to apply static standards to things in process. This is the case of the universe. We speak about the "beginning" of the universe, when the universe is a system of systems that are, have always been, and always will be, in process. There was no beginning, and there will be no end to the universe. The so-called *Kalām argument* in philosophy that "every being which begins has a cause for its beginning; now the world is a being which begins; therefore, it possesses a cause for its beginning" expresses the scientific principle of causality, which, as we have seen, is only applicable to isolated systems. But the universe is a system which cannot be isolated because we live inside it. It is a very unique system comparable to no other that we know. Since the causal conditions cannot be verified, the principle of causality does not apply to this unique system. The continuity of process is mentally broken when we think of it in terms of permanence. In other words, we can stop process mentally, but not in reality.

By being in endless process, the universe possesses timeless characteristics. Only when a system, by its own tendency to symmetry, isolates itself out from the rest of the universe does time appear and a structure, like a galaxy, can be formed which has now beginning and eventually has an end. All complex structures in the universe, physical or biological, arose from the field and ultimately must regress to the field. No structure, isolated from the field, can stay in process indefinitely. From this point of view, a structure is just a temporary organization of energy or structural asymmetry.

Unitary theory asserts that there is only one general law of nature: tendency to symmetry. All other valid laws are partial and should be explainable by unitary concepts. Furthermore, the present physical laws are seen as abstractions conceived by physicists from their top-down observations of physical phenomena. They were not derived directly from nature and so do not have empirical referents in the field. One of the consequences of this fact is that we tend to think of a force as if it were some powerful dynamic agent applied to an object, when in effect it is just a mathematical equation relating mass, velocity, and time. In reality, however, a force corresponds to the degree of the tendency of a given structure to symmetry. A force is manifested wherever there is a structure going from asymmetry to symmetry. It is structure that gives validity to our abstract concept of force.

Perhaps a more important issue, for it directly affects the future development of science, is the limitations of the present scientific method of investigation uncovered by unitary thinking. Our investigative method is based on quantity: it uses numbers and statistical methods. But there are no numbers in nature; they were put there by scientists in the process of measurement. Structure is basically relational: parts or subsystems are related to wholes, and wholes to wider wholes, by asymmetrical relations. The resultant patterns are qualitative and not accessible to quantitative measurement. Quantitative aspects do exist, but they are not determining factors in every phenomenon. In practical terms, the present scientific method leaves a lot of information behind—perhaps the most important information. Quantitative laws and statistical techniques do not yield the whole truth.

Since the discovery of the quantitative method in 1600, we have measured almost everything that needed to be measured, and further exploitation of the method or more refined measurements are not expected to lead to major advances in knowledge. But the quantitative method is not the only investigative method possible. We need to discover a new process of measurement capable of *reading* the state of order of the qualitative patterns of structure, particularly of the complex biological structure. For this to occur, a mathematical basis must be given to the present unitary model.

From the above considerations, the reader can now understand the quagmire that contemporary physics theory is facing. The challenges brought about by unitary theory are so radical that physics is reluctant even to test the theory, let alone to accept it. But the pressure is mounting, and it is just a matter of time for physicists to realize that a reevaluation of their present concept of field is long overdue.

One of the problems facing medicine now is the huge amount of seemingly unrelated data obtained by the application of the quantitative method and now waiting for some kind of a comprehensive scientific synthesis. It is a chaotic situation similar to the one going on in the economic sphere, also brought about by present economic theories based on quantity. The truth of the matter is that the quantitative method is good for analysis and for the study of isolated parts, but it is devoid of any principle of form or order, which is required for the study of organized wholes. There is thus a built-in tendency to anarchy in the quantitative approach.

It leads only to differentiation but not to organization of knowledge.

A more discriminating and incisive scientific method is needed for the study of complex field organizations, one that takes into account the whole pattern of structure without neglecting its components. To unitary theory, the whole is even more important than its constituent parts.

The delay in the acceptance of unitary theory by physicists is drastically impairing the development of the natural sciences, namely biology, psychology, and sociology. Without the existence of an endogenous field, it is not possible to understand the fast growth of a mushroom—or of a baby, for that matter. How is it possible, in so short a time, for a baby to be formed from a single cell—the fertilized ovum? Does mRNA move from the nucleus of each cell to all parts of the cytoplasm? How many mRNAs are in transit all the time? How are amino acids assembled into proteins? Is it by lock-and-key mechanism? Assuming that there are 10^{14} cells in a newborn, each with over one hundred thousand different proteins, there must be millions of cell divisions and trillions of protein multiplications *each minute.* How can this be accomplished by known mechanisms of multiplication? And without errors! It is only possible by almost instantaneous in situ protein duplications, and for this to take place, the action of a field is required.

With the application of unitary concepts to psychology by Leo Baranski, mental processes have been explained and entities such as the conscious and the unconscious clarified. Our thoughts and actions are no longer thought to be brought about by mystic mechanisms involving external agents. They are just expressions of the unitary process in a highly differentiated organic matrix, the brain-mind. In the beginning of the twenty-first century, the time is ripe for each one of us to know that a formative principle exists inside us and is constantly adapting us to our environment. It is through this process that our lives are maintained in balance.

By providing a truly scientific body of knowledge that every human can understand and believe, unitary theory should be capable of bringing humanity together. Unitary thinking can serve as a guide, not only in research but in the organization of society as well. Stable social relations and cooperation among humans can only come from mutual understanding.

Although we have discovered the process by which the universe appears to operate, we still do not know all the secrets of nature. Among those secrets are the laws of the structural transformation of structure: the laws of form. These must be approached by mathematical means. Attempts have been made in this direction, but the new mathematics still remains to be developed. Present unitary theory is just a sketch of a future *exact theory of science* that, in the vision of Whyte, should be capable of unifying all knowledge.

2

THE ORGANIC PROCESSES

It was not by fate that life appeared on the surface of our planet: a group of circumstances contributed to its development. The major factor, the one that led to the cooperative operation of the two tendencies of the unitary field, must have been the level and stability of the temperature of the galactic space around the earth. With too low a temperature, the tendency of structure to symmetry would have predominated and total symmetry reached at absolute zero; with too high a temperature, the tendency of the field would have dispersed the formed structures and negated the build-up of the complex biological edifice.

Besides allowing the cooperative operation of the two tendencies, the relatively high level of temperature on our planet kept water in the liquid state, another necessary condition for the development of organic structure. In addition, the strength of the gravitational field of earth also had to be right: high enough to hold the light water molecules evaporated from the oceans in the atmosphere, preventing them from drifting into space, and low enough to allow small drops of rain to fall. In this way, the water cycle could be maintained.

Looking at things from a different point of view, it was the miniscule portion of the total solar energy that falls on earth, the earth-sun distance, and the size and density of the earth, all combined, that created the facilitating conditions for life on our planet. Other factors then prevented the formed life from destruction. This is the case of the magnetic forces emanating from the earth's core—the *terrestrial magnetic field.* This field deflects back into space the solar wind, a continuous flow of highly energetic radiation that, if free to bombard the earth's surface directly, would have destroyed any formed living structure.

There is nothing unique about the phenomenon of life. What is truly unique is the universal formative process. Given the appropriate environmental conditions, life will inevitably develop. Although life-friendly environments very seldom occur on planets, the universality of process on one hand and the huge number of galaxies known to exist throughout the universe on the other make it likely—in fact, almost certain—that many extraterrestrial civilizations do exist.

In terms of size, the terrestrial human lies between the very big and the very small—say, between a galaxy and a bacterium. In terms of complexity, however, it sits on the top of the scale. We must consider now how this complexity was achieved and how it is sustained.

The biological field

A most embarrassing question facing present physics and biology is that concerning the acquisition, by organisms, of the properties we associate with life. The question is usually framed as follows: How can an assembly of atoms, which are devoid of living properties, give rise to animal organisms capable of walking, thinking, and talking? Obviously, without the concept of a universal formative and creative process, the life question cannot be answered.

It is true that biological systems are different from physical systems; a human being, for instance, is very different from a star. But what differentiates them, basically, is the level of organization of the energy field from where they derive—the biological field versus the physical field.

The biological field is made up of all polarized and polarizable structures of the body—mainly water, biopolymers, and ions. Since water and biopolymers are themselves molecules made up of atoms, which are structural organizations of the physical field, the biological field is no more than an extension of the physical field. It is the physical field operating at a higher structural level of organization within a circumscribed portion of space.

Most biopolymers are proteins, macromolecular structures composed of various proportions of about twenty commonly occurring amino acids, themselves made up of atoms of hydrogen, carbon, oxygen, nitrogen, and sometimes sulphur. All these atoms possess residual asymmetries and latent symmetries of their own, but these are still obscure.

The hydrogen atom is the fundamental unit of asymmetry or polarization and tends to complete the symmetry of more complex patterns; the carbon atom has four central asymmetries that can be complemented by combining with other carbon or hydrogen atoms into symmetrical chain or ring molecules.

In nature, one carbon atom can bond with four others to form extended three-dimensional networks. And, under high-pressure conditions, a carbon network of this kind can crystallize into the mineral known as diamond. The diamond is the hardest substance known, and it owes its hardness to the strength of the carbon-carbon bond. This ability of carbon to form durable and stable compounds was exploited by nature for the build-up of the organic structure.

To facilitate description, we can compare the skeleton of an amino acid to the shape of a cross: two intersecting bars of unequal length. Amino acids are so called because they possess both an amino group, which contains nitrogen and hydrogen atoms, and an acid group, which contains carbon, oxygen, and hydrogen atoms. These groups are attached to the same carbon atom, known as *α-carbon*. In our descriptive model, the α-carbon sits on the intersection of the two bars, with the amino group on one side and the acid group on the other, on the horizontal bar. On the shorter portion of the vertical bar lies one hydrogen atom attached to the central α-carbon, to which is also attached a variable residue lying on the longer portion of the same bar. This residue, which can be just one hydrogen atom or a complex structure such as a heterocyclic ring, characterizes the specific amino acid.

In a typical protein, about one hundred to three hundred amino acids are joined together by their shorter bars, forming carbon-carbon and carbon-hydrogen bonds, the acid group of one attached to the amino group of the next, and so on, and so on, to form a characteristic sequence or main chain. The vertical bars of the crosses, in different orientations, form now the side chains. The main chain is then folded over upon itself many times, in a very specific manner, to finally reach its stable three-dimensional shape. Most proteins are globular, having some resemblance to the shape of potatoes, but many other shapes exist, dictated by their special functions.

When a protein is first formed, some of the asymmetries between a side-chain amino group in one fold and a side-chain acid group in a neighboring fold, belonging to amino acids located far apart in the extended main chain, are complemented by cross-connections, such as hydrogen bonds (N–H–O). These special bonds are much weaker than the strong chemical bonds of the main chain. Nevertheless, they still contribute to the stability of the molecular framework, while lending some flexibility to it. It is the flexibility of these linkages that is responsible for the constant motion of a protein.

From the unitary point of view, a protein combines in a unique manner the properties of stability and instability, or of symmetry and asymmetry. The tendency to symmetry is clearly expressed by its capacity to form a linearly symmetrical chain—the main chain; the tendency to instability arises from the asymmetrical (polar or polarizable) side and end groups strung on the above chain. These groups are arranged regularly along the molecule, thus establishing a continuous resultant field that extends to, and even beyond, the confines of a protein. Seeing it this way, a protein is just a relatively stable carrier of specific polarization.

Functionally, a protein cannot be described in isolation from its surrounding water and ions. All these components form a single system—a living protein. The highly asymmetrical and flexible water molecules are capable of taking up and propagating the polarization fields of proteins, though at great distances, measured in micrometers, this tendency is rendered ineffective by heat motions. And the polarized ions are agents of change and also carriers of induced polarization.

Each living protein is itself surrounded by hundreds of smaller molecules of more than ten different kinds. The resultant protein field influences the orientation of all these adjacent structures, involving them in a particular function. All together they form a *functional system.* This functional interdependence is of course lost if a specific protein is taken out of the body.

It is protein systems that constitute the basis of the functional patterns of animal organisms and determine their general properties. All other non-protein constituents serve some special role in relation to the protein processes. But a functional protein system in a given region does not operate independently. It is itself a component of another, wider system, and this interrelation and interdependence ultimately involves the whole organism. The whole body can be viewed as a single system. It represents, in effect, the whole biological field.

The two tendencies of the physical field are also manifested by the biological field. Proteins, as both structures and components of the biological field, usually possess two characteristic states of polarization, which are separated by a relatively low threshold. Under undisturbed conditions, they are held at the higher polarized state by the resultant field of the region. When the protein molecule is disturbed by an external or internal influence, it will tend to pass into the more symmetrical depolarized state, thus expressing the natural tendency of any structure to symmetry. The restoration of the previous asymmetry, the function corresponding to the normalizing process of the physical field as a whole, is then accomplished by the wider system of which the protein is a part. As a consequence of the alternating depolarization and polarization processes, a living protein is constantly pulsating.

The polarizing cycle of one region is always causally linked to the cycle of the next. In this way, only one normalizing process finally dominates the whole organism. Alternating pulses of depolarization and polarization are then propagated without interruption along channels of identical or cooperative molecules.

In the unitary view, all processes in organisms are *directed processes;* that is, all processes occur not only at particular times and places but in particular orientations, all largely determined by their relation to the protein framework. The directed character of the protein processes is, however, constantly challenged by external and internal stimuli and by thermal disorder.

The regularly oriented patterns of functional protein tend to collapse if proteins are excessively disturbed. Protein systems, therefore, must have a high degree of polarizability; that is, they must have high polarization fields relative to the size of the molecule. This is the reason why proteins are large molecules or extended networks, possessing regular and sufficiently close repetitions of polarized groups. In addition, they must function in a medium, such as water, that allows adequate scope for intra- and intermolecular changes and helps to maintain the polar groups in a state of polarization. Under these circumstances, and if the temperature is not too high, protein structures and their cyclic transformations tend to become stabilized. There is a bias on the side of the ordered, as opposed to the disordered, processes.

A closer look at the biological field reveals that structure, or particle, cannot be differentiated from field. The process of the field consists of changes of the state of polarization and relative positions and orientations of the parts that constitute a structure. Structure and field are indistinguishable. It is this fact that underlies the unitary concept of a structured field.

The unitary biological field includes the electric and magnetic fields of classical physics. In unitary theory, an *electrical charge* represents a polar asymmetry. And an *electrical current* occurring in living structure is considered to be a continued induction, by a chemical source, of an extended state of polarization.

The normalizing of the field

Like its physical counterpart, the biological field possesses a tendency toward uniformity of asymmetry, which is achieved by inducing changes of shape, orientation, and position in polarizable structures. In this way, the process of the field—the normalizing process—brings each polarizable part of a given system into conformity with the resultant polarization of the whole system. This occurs in all systems throughout the whole organism. A compromised state of polarization, a state of minimal polarization differences, is thus established in the biological field. This is the normal state of polarization, or *norm.*

The norm, therefore, is not a single value but instead a range of values. In other words, there are as many different norms as there are different systems in the living organism. And each of these norms is not a static condition but instead a pattern of processes, each with its normal rhythm in accordance with the environmental situation; i.e., the season, the hour, and the current activity. The norm of an organism is thus very sensitive to variations of its environment.

The normalizing process of the biological field is not basically different from that of the physical field. The biological laws are the same as the physical laws, to be sure. But due to the extreme structural complexity of organisms, the biological field alone cannot sustain the living process. It requires extra amounts of structural asymmetry or free-energy.

The need for additional free-energy was felt very early in evolution, during the development of the first unicellular organism. It was precipitated by the increasing complexity of the evolving cellular structure. To promote its orderly development in the face of constant disruption by the external environment, the cellular organism needed a stable medium that could only be achieved by the packing of a great number of structural components into a limited space. To that effect, water molecules bonded to each other by hydrogen bonds to form clusters, and these water clusters, together with ions, were then adsorbed onto the external surface of proteins and other "solid" structures, ultimately leading to the formation of an immobile gel. But the energetic cost of raising and sustaining the polarization level of the concentrated water-protein gel proved to be too high.

When the asymmetry level of the physical field was reached, further cellular development became possible only at the expense of the organism's norm. With the norm declining and threatening to decrease polarization to a level incompatible with life, a true energy crisis arose. These events occurred over three billion years ago.

The energy crisis of evolutionary development bears similarities to the present energy crisis of social development: both were brought about by increases in complexity. Our increasingly sophisticated civilized life is requiring more and more free-energy for sociocultural development to proceed unimpaired.

Nature solved its energy crisis through the development of cellular metabolism, which extracts structural asymmetry from an outside source. Primitive unicellular organisms first used the electronic energy of chemical compounds readily available in the outside medium, and only later in the evolutionary process was the nuclear energy of the sun utilized as a definite source of free-energy. And it took millions of years more for the normalizing process of the cell to develop cyclic metabolic structures similar to those we have today.

The development of present-day metabolism involved the close cooperation of bacteria, plants, and animals, both in the sea and on land. Cyanobacteria, a type of phytoplankton, were among the first organisms to use the free-energy of sunlight to fix the carbon atom of atmospheric CO_2 into carbohydrates (sugars) in the process of photosynthesis, with release of the oxygen molecule into the atmosphere. With photoenergy, those bacteria and others are also capable of fixing atmospheric nitrogen in the form of ammonia or nitrate ions, which are then assimilated from the soil by plants for the synthesis of amino acids and proteins. Later in evolution, the photosynthetic process continued in plants possessing the green photon-capturing pigment chlorophyll, which enabled them to synthesize carbohydrates and fats. Among these green plants are the various types of grasses, the foodstuffs of herbivorous animals that, in turn, supply food for the carnivores. All our foodstuffs—fish, meat, fungi, plant roots, cereals, fruits, and vegetables—are thus ultimately assembled by the energy supplied by our sun.

One organic compound vital to the initiation of the metabolic process was *adenosine triphosphate* (ATP). This complex molecule is capable of transporting structural asymmetry from one molecular system to another without using it or dissipating it in the form of heat. It consists of adenine (a nitrogen-carbon-hydrogen compound) linked to ribose (a carbon-hydrogen-oxygen compound), which bears three phosphate groups, linearly linked together by strong bonds. The free-energy is temporarily "stored" in these phosphate groups.

In unitary biology, the synchronized processes of anabolism and catabolism correspond to the two tendencies of the field—tendency to symmetry and tendency to restoration of asymmetry—working in close cooperation in living structure. Anabolism builds up complex and more symmetrical structure, and catabolism supplies that structure with the asymmetry required to bring it to the asymmetry level of the field, thus preventing the formed structure from sliding into complete symmetry.

The ATP system is the structural unit that serves the normalizing process on the molecular level. In present life, your body synthesizes approximately its own weight in ATP daily, which acts as the energy source for cells. In the unitary view, ATP is an agent of the field; the structural asymmetry of its terminal phosphate groups is actually a part of the normalizing process itself.

With the sun as an inexhaustible source of structural asymmetry, the field process could now continue with further development of the biological structure, including the process of reproduction. But the norm of the evolving organism was becoming ever more sensitive to the distortions induced by the environment. The evolutionary process would be much accelerated if the fragile organism could somehow develop the capacity to withstand environmental insults without using too much energy.

Facilitation of the normalizing process

Every formative process possesses a threshold. Whenever this threshold is reached in a given system, by a stimulus that is not too excessive, its structural pattern spontaneously slides toward symmetry. This event triggers a normalizing pulse in the wider system of which the given system is a part, which stops the sliding to symmetry and brings the now transformed, more symmetrical pattern of the part back to the asymmetry level of the whole. The normalizing pulse carries structural asymmetry.

Stimuli lower the asymmetry level of the normalizing process and may therefore be considered normalizing distortions. But stimulation within a certain range is essential for the maintenance of life. Organisms are sustained in process by the continuing stimulation of their environments.

Early in evolution, the primitive unicellular organism was very sensitive to environmental stimuli. Changes in pressure, temperature, electromagnetic radiation, gravity, and internal chemical variations of ionic and molecular species all tended to distort the normalizing process, with great expenditure of structural asymmetry to sustain the norm.

Under these circumstances, what could the evolving organism do to avoid or minimize the requirements of structural asymmetry, which at this time was being supplied almost entirely by the physical field? Judging by the course of later events, it appeared that the only way out was for the organism to slowly adapt to its environment.

But how could such a feat be achieved and sustained in the long run? How could the structural patterns of the organism be made similar to the patterns of the variables of the environment? The normalizing process finally solved this conundrum by developing a special structure where the structural patterns of the environment could be recorded and constantly updated. This structure could then direct the buildup of all protein structures required for the functioning of the cell. And slowly, the ATP molecule constructed, from available nucleotides, the long chain of DNA. The DNA became the inner boundary of the normalizing process of the cell, the outer boundary being the cell membrane, which constituted the interface between the external and the internal environments.

Unitary theory postulates that the stimuli are made up of patterns of quantum field particles, which are aggregations of free-energy particles modified by the environment. Within the organism, the quantum field particles formed quantum structural aggregates, smaller in size than any nuclear particle. These structural aggregates can be identified with what are now known as *genetic memories*. Presumably, they are scattered throughout the nucleoprotein nuclei of the chromosomes.

Because of the transactional nature of the formative process, what comes to DNA combines with what is already there, and so no past information is lost. When the variables or conditions of the environment change, the normalizing process makes compensatory changes in the genetic memories—by selectively inhibiting, reordering, or creating them. It was this adaptive characteristic of the formative process that made one-way development possible.

After countless cycles between cell membrane and DNA, the normalizing process, through its ATP agent, slowly built up protein enzymes from available amino acids. The enzymes then developed structural and functional organizations, which led to the appearance of the first living unicellular entities.

Over eons of time, and with cell duplications, highly ordered multicellular organisms made their appearance. Structural and functional organizations were developed on the large scale of size—metabolic organs and their connections, sense organs, glands, limbs—which facilitated normalization on a higher organizational level. Throughout the one-way development of the animal kingdom, genetic memories were the fundamental units of biological order.

With the passage of time, the organic structures became increasingly more symmetrical until a workable level of steady symmetry was reached. The more symmetrical structures were more stable and therefore deviated less from the norm than the more asymmetrical ones. In this way, the symmetrical structures also facilitated the attainment of the level of uniformity of asymmetry, or norm.

But the tendency to symmetry is an inherent property of structure, while the process of adaptation to the environment is a property acquired by organisms and without which they could not have survived.

Water as basis of the organic processes

Chemical processes, as any others, involve structural transformations or changes in the geometrical relations of structure. So, to properly understand the structural transformations undergone by a given chemical system, it is necessary to know the latent asymmetries and symmetries of all its components and, ultimately, of all organic atoms. However, our present atomic theory is mechanical in nature and cannot supply such knowledge. This limitation and many others prevent us of giving a coherent unitary description of even the basic enzymatic reactions, let alone the myriad of structural transformations that occur in the body.

Nevertheless, something is already known about the geometry of the most fundamental organic atoms—hydrogen, carbon, and oxygen—and we can use this limited knowledge to give the reader a better understanding of the profundity of unitary theory.

Polarization fields are directional. Therefore, to facilitate description, we can take a structural asymmetry to be a vector quantity. Classical field theory had already recognized the polarizing abilities of the oxygen atom, but it was unitary field theory that clarified them. This atom contains two asymmetries; that is, it is missing two electrons to complete its own symmetry. These asymmetries correspond to two oblique vectors, and this means that it is not easy to bring the oxygen atom into complete symmetry.

Usually, when we talk about symmetry, we equate it with stability, asymmetry being related to instability. But here we have an example of "stable instability." And it is from its inherent stable instability that the oxygen atom and most oxygen compounds derive their special properties: the oxygen field is capable of inducing asymmetry in symmetrical but less stable molecules, such as the fuel molecules, by ripping off electrons from their structures. This is the process of oxidation, in unitary terms.

Let's now consider the case of water, the molecule that is at the basis of the living process. We know that the water molecule is composed of three atoms, one of oxygen and two of hydrogen. The reason for two hydrogen atoms and no more is that the oxygen atom is missing only two electrons, which the hydrogen atoms are trying to replace. Under these circumstances, and given the rather rigid oblique orientation of the asymmetric vectors of oxygen, the water molecule must necessarily be triangular in shape: it is a triangle without one of its sides. On each open end of the two sides sits one hydrogen atom, and the oxygen atom is located where the sites meet. Since the attraction of the electrons of hydrogen by oxygen is stronger than that of hydrogen itself, the shared electrons spend more time closer to the oxygen atom. In present biochemistry, this asymmetric distribution of charge is called a *dipole,* but in unitary biochemistry it is just a specific polar asymmetry, or polarization. The water dipole is always present—water is a permanent dipole.

In the water molecule, the oxygen atom is not completely stabilized by the two opponent hydrogen atoms. We can look at this hydrogen-oxygen relationship as two lightweights trying, in awkward positions, to overcome one heavyweight in a wrestling match that goes on indefinitely: the powerful oxygen atom can only be tamed, at best.

In liquid water, all the hydrogen and oxygen atoms are involved in the match, the arms of the wrestlers being entangled through dipolar-dipolar interactions called hydrogen bonds. This molecular arrangement leads to the formation of a continuous, three-dimensional, and flexible hydrogen-bonded network. Liquid water is thus in constant change, and it is in change that the secret of life resides.

As a liquid, water is unique. In effect, it is supposed to be a gas at ambient temperature, but the mutual hydrogen bonding between water molecules does not allow them to easily break free and vaporize. Unlike a normal liquid, which increases its density as the temperature decreases, liquid water does the opposite: it decreases its density upon freezing, by stretching the hydrogen bonds, and so ice floats in lakes and ponds, permitting fish and aquatic mammals to continue living underneath.

The molecular instability is spread throughout the continuous water network. One of its consequences is the experimental finding that liquid water is composed of a mixture of two constantly interchanging populations of differing densities—low-density water (LDW) and high-density water (HDW)—dependent on the degree of tightness of their constituent hydrogen bonds. LDW is characterized by straight and strong and HDW by extremely bent and weak bonds, surrounded by many unbonded molecules. Conceivably, a continuum spectrum lies between those extremes.

Bulk water is continually oscillating between structuring and destructuring, groups of water molecules constantly going from LDW to HDW and back, from ice-like to vapor-like water, from a solid to a gas state, so to speak. The velocity of these events is far beyond what our eyes can see, and so water appears to us very stable.

But the liquid environment within cells is even more complex. It contains dissolved solute, mostly ions (polarized atoms or groups of atoms) and proteins. Protein surfaces have charged and uncharged moieties. The charged moieties attract counterions, so that in a layer of about 2 nm in thickness around a protein surface there is a high concentration of ions. As water changes its structure, its solubility properties change too, so some ions are attracted to LDW and others to HDW. This generates local structural asymmetries that are involved in bioenergetics. Water is not only in constant change but also tends to promote change.

Up until now we have been dealing with the normalizing process that involves the whole organism, which is made up mostly by the summation of the protein fields. But within each protein field there are numerous nanofields emanating from the polarized and polarizable structures it contains—ions, some small solutes, and water molecules themselves. These small fields are particularly important near the surface and inside proteins, in the narrow clefts between domains of the folded structure. It is in these small cavities that most, if not all, active sites are located. And it is from these interacting fields and processes that the enzymatic reactions that support life result.

Development of organisms

From all available theories, only unitary theory gives us a scientific understanding of the origin and development of organisms. It postulates that the universe is a field of structural asymmetry and that all existing structure—inorganic, organic, and mental—is ultimately derived from the field. Once separated from the field, structure increases in complexity by a one-way, or directed, formative process of decreasing asymmetry. More complex symmetrical forms are necessarily developed from simpler but more asymmetrical ones by a series of structural transformations, the sum total of which constitutes what we call evolution.

The term evolution, however, is misleading. It comes from the Latin *evolutionem,* the unrolling of a scroll. Thus, the term contains the assumption that the development of life on earth represents the unfolding of a coherent plan aimed at a predetermined goal. Nothing could be farther from the truth. Life processes are not determined by future events, and the order existent in nature is not a coherent one. There is always a component of disorder in natural processes, which we associate with chance. We live in a world of natural disasters and of accidents brought about by human action.

The highly complex human organism, therefore, could not have been built in one step. Its structural arrangement required the successive transformations, through countless generations, of simpler forms of life. The agent directing and organizing this development was the normalizing process of the field, the force behind existence.

The materials for the buildup of living structure must have been chosen by the field from the surrounding environment and then worked by the field. It was not by chance that only atoms of low atomic number (number of protons in the nucleus), and thus low number of orbiting electrons (i.e., H1, C6, N7, O8), were selected for the job. In these atoms, the electrons are tightly bound to the nucleus, and there are few free (conducting) electrons. An electrical field applied to these atoms causes a displacement, but not a flow, of charge. The electrons are displaced by the applied field with respect to the nucleus, giving rise to an induced dipole that has an electric dipole moment in the direction of the field. Bioatoms and molecules are poor conductors but easily polarizable structures, very responsive to the biological field.

Besides being constructed from polarizable molecules extended in space, living structure is generally neither rigid nor excessively loose. Its components are allowed to have some movement so that they can adjust to the orientation imposed by regional fields and normalizing pulses. These characteristics are akin to those displayed by liquid crystals, crystals with some molecular mobility.

The so-called liquid crystals are substances that flow as liquids but maintain an ordered structure as crystals do. Strictly speaking, however, they are not crystals and need not be liquids. Rather, they are systems whose constituents possess long-range orientational order but not positional order. A more apt name would be *orientationally ordered systems.*

A liquid crystal has some characteristics of a crystal and some of a liquid, but generally it has unique properties of its own. Because the molecular forces underlying the liquid crystalline state are very weak, these structures are easily affected by changes in mechanical stress, electrical and magnetic fields, temperature, and chemical environment. All these properties are ideal for organisms.

Many liquid crystals consist of elongated, rod-shaped molecules, which for energetic reasons tend to line up in parallel. It is intuitive that densely packed rods can move around more freely if they are aligned. Similar characteristics are seen in some biological liquid crystals, such as the lipid bilayer that serves as cell membrane.

But in the body, liquid crystals are not limited to cell membranes. They are present everywhere, inside and outside cells. Within cells, most of the proteins, DNA, and the cytoskeleton constitute liquid crystal systems, and without cells, the collagen fibrils and proteoglycans of connective tissue also are liquid crystals. It can be said that all functional parts of organisms are made up of liquid crystals.

Furthermore, there exists extensive coordination between the positions and orientations of functional systems in an organism. And within each functional system, the parts are related to the whole in a very specific manner, determined by their states of polarization. The structural pattern of the whole, being more symmetrical and more stable, is dominant in its relations to the more asymmetrical and less stable structural patterns of the parts. The parts tend to conform to the whole in developing its symmetry. They appear to cooperate with the whole in this process. It is cooperation and not competition that characterizes formative processes. Regrettably, the beauty of this orderly arrangement escapes present biology, for it misses the asymmetrical relations of dominance and subordination.

The functional unit of all organisms is the cell. This is so because only this entity is capable of reproducing its own organization, by duplicating its components. Cells are very complex structures, even those of unicellular organisms, such as bacteria.

The complexity of a cell is thought to result from the great number of constituent parts and the multiplicity of connections between them. But some of this complexity may result from the outside-in perspective of present biology. The postulated intermolecular connections disappear from consideration when the relation of wholes to parts of unitary theory, based on the intrinsic polarizing properties of structure, is taken into account. The extreme cellular complexity may thus be more apparent than real.

If unicellular organism is complexity, multicellular organism is supercomplexity. Higher organisms, made of interconnected tissues and organs of different types of cells, where countless processes operate, display massive complexity. But complexity is not infinite, even in thinking organisms, where it reaches the highest degree. It is limited by the relatively low number of different atoms and the few tissues and organs that make up the organic structure. It is thus accessible to the human intellect.

Moreover, in a cell, and in the organism as a whole, the adding on of parts is a self-limiting process. When the structural pattern of a given system reaches a certain degree of complexity, the system's behavior becomes unpredictable and chaotic, and no further buildup of structure is possible. A new structural pattern must be started de novo. Thus, biological organization is necessarily hierarchical in nature.

For proper functioning of an organism, high complexity must be matched by high coordination. The complex human organism therefore tends to be highly coordinated. If running is initiated and more work is done, more energy is required, and this means that all the metabolic organs must work together to supply it. The heart pumps faster, the lungs increase their respiratory cycle, and the liver augments carbohydrate metabolism so that enough oxygen and fuel is delivered to the working muscles. All of this is done under the automatic supervision of the nervous system, which may also increase its attention process.

This internal coordination must have been established and perfected during evolutionary development, along with the process of adaptation to the external environment. Having been generated by the universal formative process, the coordinated tendency is therefore an expression of the ordering process of nature. In a constrained system of mutually dependent organs, not every change is possible. The change in one organ is conditioned by the change of all related organs, in the cooperative performance of a single function.

In present life, the entire complexity of the adult organism must be contained in the hereditary units (genes), in the nucleus and cytoplasm of the fertilized ovum. The hereditary units are extremely stable, and their structural patterns are dominant. The development of the organism consists, at bottom, in the extension of these dominant structural patterns by the normalizing field.

Organic stability and self-regulation

One of the remarkable characteristics of organisms is their manifested stability. Our bodies do not lose their forms overnight, and our muscles always return to their previous shapes at the end of contraction and do not simply fall apart. Some parts move about in the course of growth and function, but the underlying order is undisturbed. The body structure is persistent.

But the organic stability is limited and can be dispersed by a stimulus exceeding a given threshold. It is a threshold-limited stability. Basically, it is an expression of the structural stability of our protein molecules, or more specifically, of their rigid, linear, symmetrical chains.

The constituent proteins of our body reached their symmetry long ago, in the course of the eonic evolutionary process of development. They are now stable structures, but their symmetry is not complete: it is distorted to conform to the system of the organism.

Furthermore, the symmetry level is not identical in all structures. Some, such as genes and certain protein enzymes, have reached a high degree of stability. Their internal structure is no longer changing—they are static. The structural state of these proteins is comparable to that of the solid crystals of the inorganic realm. They are, however, still capable of extending their form by duplication, through the action of the normalizing field. In effect, it is only in this highly developed symmetrical state that proteins can be replicated. And, as previously stated, it is only to this limiting case of steady symmetry that the unitary term "structure" is properly applied.

There are other stable structures, such as the proteins of contractile tissue, which are not static. They undergo a minor reversible change in the course of their function—they are cyclic. These cyclic structures correspond to the liquid crystals previously described. But during the cycle (i.e., folding and unfolding), it is only the internal structure that is transformed and not the external form. The latter remains unchanged and therefore stable. Furthermore, the internal cycles themselves, when constantly repeated, tend to become stabilized, in the sense that their thresholds tend to diminish, permitting the cycles to be more easily performed. In the special case of muscle tissue, the contractile proteins are arranged into fibers through which a polarization pulse propagates. Under these conditions, the propagation of polarization results in a coordinated change of molecular shape, which ultimately leads to macroscopic contraction.

Besides structural stability, organisms display another, equally remarkable property—that of self-regulation, or the automatic return to the norm after disturbance. Any deviation from the norm always evokes the function that eliminates that deviation.

But in what possible way can such magic behavior be achieved?

It happens that during the course of function, be it duplication or cycling, structure responds to normalizing fields with characteristics similar to the field by which the structure was originally formed. Self-regulation is thus directly related to differential synthesis. Any structural pattern appears to carry with it a "record" of the process by which it was synthesized, and this record evokes a normalizing field of the same character any time a structure deviates from its norm. Self-normalization is thus the most striking example of facilitation of the normalizing process.

Organisms work the way they do because they are constructed in a specific manner. The automatism of living structure is based on the structure-field relationship and on the characteristics of the biological process itself. Unitary theory has discovered one of the deepest secrets underlying life—its automatic functioning.

3

THE MENTAL PROCESSES

One of the most debated subjects since the beginning of civilization has been the nature of, and the relation between, what we call body and mind. The nature of the body has always been taken to be physical, but the mind does not appear to be made of any known physical substance. What the mind is made of has always escaped human comprehension.

Initially, when the term was coined by the early philosophers, "mind" may have referred to human thought; but over the course of time, its meaning expanded to include the faculties of consciousness, volition, and feeling. The mind was considered an entity in itself, which was localized to the brain but separated from it.

Not only was the mind separated from the body, but it was also physically separated from the rest of nature. By necessity, therefore, the mind was assumed to be controlled by some ill-defined spiritual agency or divine principle. This kind of thinking dominates religion and philosophy, and it also dominated the first centuries of science.

But a deeper insight into the body-mind problem reveals that what differentiates mind from body is the mind's dependency on time. The memory of past events and the anticipation of future ones are clearly time-related. The body, however, appears to be independent of time and indifferent to it. Time is the essential mark of process.

As discussed previously, the body structures have reached maximally possible, but not complete, symmetry during the one-way process of development. They can still extend their form when necessary and are under the normalizing process of the organism. They are isolated but not separated from process. In effect, they are prevented by the norm from reaching further symmetrization. The mental structures, however, continue to be in a continuous process of transformation. In other words, the body structures display only the spatial aspect, while the mind structures continue to display both the spatial and temporal aspects of process.

In the unitary interpretation, mind and body do not exist independently. These terms just refer to two phases or aspects of the same organic process of development. They are not a pair of contrasted aspects. There is thus no body-mind dualism. Dualisms result from the intellectual breakup of the continuity of process, but one cannot divide what is continuous in nature.

Body-mind dualism is the child of static thought. In ancient times, when the thinking process was evolving, the human could not possibly conceive of process. He had been thinking and communicating in static terms all along and could only comprehend static ideas. Perhaps he even needed static ideas to consolidate further his nascent knowledge. A process running in nature and man was recognized early on in Greek philosophy, but its concept was too vague for proper understanding. Unfortunately, static concepts, such as the sharp division of body and mind, are still with us today, standing as a barrier to the general acceptance of unitary thinking.

There is also no separation of organism from nature itself. The universal formative process runs through all realms, inorganic and organic. Furthermore, an organism is under the constant developmental influence of its environment, from which it cannot be separated. Environment and organism form a single system.

In unitary theory, a system is composed of polarized parts held together by ordered relations of asymmetry, each system being itself a part of a wider system. When this interrelation and interdependence reaches the level of the whole organism, the organism system becomes a part of the wider system of its environment—the organism-environment system.

The organism-environment system

Organisms do not evolve spontaneously. They are the end result of a series of chemical transformations of structure brought about by metabolic processes over a vast expanse of time, under the guidance of the normalizing process.

The basic unit of the organism, where the normalizing process is at work, is the cell. The normalizing process extends from the cell membrane, the outer boundary, to the hereditary units, the inner boundary. But while the inner boundary is fixed, the outer boundary is variable and under the control of the environment.

For life to be sustained, the variability of the environment must be within an upper limit and a lower limit. The environment must at all times stimulate the organism, but this necessary stimulation must not be too variable or too constant. The organism, in turn, is constantly trying to adapt to its environment, but complete adaptation is never reached: an imperfection of fit always persists. It is the difference between organism and environment, together with the perpetual tendency of the organism to decrease this difference, that makes the organism work.

The environment supplies the organism with structural asymmetry in the form of food (aggregates of molecular structures) and in the form of stimuli (aggregates of quantum field structures). The processes of the organism are thus sustained by the processes of the environment, but it is the normalizing process, working in the organism, that ultimately controls the organism's differentiated development and adjusts it to the available potentialities of its environment.

During most of the evolutionary process, the development of the human species was slow, and its adaptation to the environment was accomplished by gradual changes in the hereditary constitution. But about ten thousand years ago, and particularly after the establishment of urban life, human intellectual development became relatively fast, and a new kind of environment—the social environment—emerged. Since then, intellectual development has been due to a cumulative modification of the social environment by each successive generation, rather than to evolutionary changes in the hereditary constitution.

It took several billion years for the normalizing process to develop the links and processes that mutually connect humans to their environment, but some processes are still developing today. They are all part of the central nervous system, a highly complex structural organization located in the brain.

The above processes developed gradually, and their increasing complexity forced the establishment of at least three interconnected hierarchical levels of structure—tropistic, perceptual, and cognitive.

In a simplified overview, each hierarchical level contains memory areas, where the quantum field structures of the stimuli are stored, and a structural center, where the formative process is at work. There are two different kinds of memory areas: sensory and motor. The sensory memories, crossed by the afferent stimuli, are located before the formative center, and the motor memories, crossed by the efferent stimuli, are located after it. The number of sensory-motor memories increases with the hierarchical level—one in the tropistic, two in the perceptual, and three in the cognitive.

While the memory organizations are located in separate areas of the brain, the formative structuring centers are located in the reticular formation of the brainstem. Presumably, the architecture of the reticular nervous structure facilitates temporary isolation of stimuli from the environment, thus allowing the unitary formative processes to go on.

In psychological processes, as in genetic ones, aggregates of quantum field particles, and not atoms and molecules, are the common denominators of structure. In other words, the nature of all our perceptions, thoughts, and actions is physical, not chemical. In a way, the physical field, modified by the local environment, extends into the organism. The biochemical atoms and molecules would not be flexible enough to undergo the fast series of structural transformations at the memory areas and formative centers, which underlie our perceptual, cognitive, and motor processes.

All environmental stimuli consist of quantum field particles arranged in a specific pattern or configuration and endowed with a certain level of structural asymmetry and symmetry. To reach the highest cognitive formative center, the stimuli must travel through all sensory memories and centers below. On their way up, and at all hierarchical levels, they decouple and bring with them memory organizations of identical configurations previously stored in the memory areas. On reaching the respective formative center, all these structures are transformed into complex quantum structural aggregates. The decoupled memory organizations are immediately replaced by identical copies.

From each center, the formed structural aggregates can follow three different paths: upward to the next hierarchical level, backward to the sensory memory areas just crossed, and to the motor areas of the same level. When the cognitive center is reached, and with no further hierarchical levels to cross, it is postulated that the upward path, now carrying very long sequences of complex structural aggregates, is here converted into an extended feedback loop that connects and synchronizes all memory areas and formative centers of all hierarchies below, before returning to the cognitive center. This continuously operating common asymmetry chain, called the *central process,* represents the normalizing process at its highest level.

Looking at the psychological organization as a whole, we have a hierarchy not of scales but of formative processes—sensory and motor. The sensory processes connect the environment to our central nervous system, and the motor processes connect our central nervous system to the environment.

The construction of our perceptions

In order to stimulate the organism, the environmental variables must introduce some normalizing distortion. The stimuli must be strong enough to overcome the threshold for depolarization of structure and must also possess extra structural asymmetry, which the organism uses as source of information about the environment, for purposes of survival. All other unnecessary stimulation should not be allowed to enter the organism, where it could severely disrupt the norm. The minimal-maximal range of stimulation is established by the normalizing process at the level of the sense organs of each species.

As an example, we take here the case of the human visual system, because the development of the eye, which became our most important sense organ, has been a challenging issue for non-unitary theories.

In the course of the evolutionary process, the sun's rays introduced progressively more normalizing distortion into the growing organism. To drastically decrease the amount of distortion introduced by this unavoidable stimulus, sunlight was allowed to enter into the evolving organism only through two tiny orifices, later to become the pupils. The photonic wave range of stimulation was then defined by the electronic structure of a characteristic protein at the level of the retina, the multilayered neural surface lining the back of the eye. In the case of the human species, the ideal wavelength range turned out to be 700–400 nanometers (nm).

But the solar photons are made of aggregates of free-energy particles in the form of electromagnetic radiation, and not of what we call light, the clarity that makes things visible. If this is so, how is light generated, and how does it relate to the received photons? It must be something produced inside the brain!

The relation of solar radiant energy to light is to be found in color—we see objects through color. Color is a *sensation* ultimately produced in the perceptual center of our nervous system, where the formative process takes place. It does not exist in the real world.

Color results from transformations of the patterns of quantum field structures emanating from the retina and complexed along the way by patterns of similar configuration previously stored in the tropistic and perceptual memories. It is in the perceptual formative center, at the moment of symmetrization and induction of structural asymmetry by the normalizing process, that we perceive (see) the objects, events, or situations in space.

The selected 300 nm wavelength range of solar radiation—from 700 to 400 nm—is viewed as a continuously varying range of colors, but it is usually split into just seven color components: red, orange, yellow, green, blue, indigo, and violet—the colors of the rainbow. This splitting or dispersion of light, in wavelength segments of ten or more nanometers, is called the *light spectrum.*

When all the components of sunlight enter our eyes, we see white light; if the eyes receive isolated sunlight spectral components, we see the color correspondent to their narrow wavelength; and when we receive a mixture of wavelength components, we see the hue correspondent to their combination. Thus, a traffic light is red because it emits red photons, those encompassing the range of about 740–620 nm wavelength.

But plants, animals, and most colored man-made objects owe their color to pigments or dyes. In this case, these substances absorb some components of white light and reflect the rest. For example, human blood looks red because the pigment hemoglobin, contained in the red blood cells, reflects light in the red part and absorbs it in the yellow and violet parts of the spectrum. And the fashionable blue jeans, which contain the dye indigo, look bluish-red to us because this pigment absorbs the yellow-green part of the spectrum. If all components of visible light were absorbed by a material, it would look black to us.

The fact that color is a subjective phenomenon does not diminish the beauty of a red rose. The red is not in the rose; it is in us. But we would not be able to see the rose, or any other object in the real world, if were not by way of some color. Without color, the universe around us would not be black. It would be invisible.

All other sense organs were developed to diminish the normalizing distortion introduced by their respective stimuli, and each one has its own range of stimulation. Presumably, the incoming quantum field patterns of the stimuli to the sensory memories and the outgoing quantum structural aggregates to the skeletal muscles travel within the nuclei of the nervous pathways.

The nature of our thoughts

We may begin with the definition of thought given by Aristotle in the fourth century: "Thought is what the soul does." Like most other definitions, it tells us nothing about thought. It just replaces one word for another that requires even deeper explanation.

Two millennia later, Descartes asserted that "thought is the unequivocal evidence of one's existence." Throughout this time period, there was no real advance in the state of knowledge in this area. The development of form in nature and of the origin of ideas in the mind was still not understood.

As long as matter and mind were considered two independent substances or modes of being, something was required to bring them together—the concepts of soul and existence. Implicit in these concepts there is an element of continuity: a rudimentary equivalent of process.

In more recent days, with the advent of computer science, a new conceptual bridge between matter and mind was advanced—that of *information*. For Eric Baum, an American artificial intelligence researcher, "Thought is a computer program," one that extracts meaning from complex data. These informational patterns, however, are not intrinsic to structure, and therefore structure and mind are not linked in a fundamental physical way.

Classical quantum field theory finally introduced the mind back into physics by admitting that the mental states are intrinsically inner states caused by neuronal processes realized in neuronal systems. But in this theory there is still no physical continuity between the processes of nature and those of the mind. As a result, there is no organization between the different psychological processes, the memory's structures are not well defined, and the process of thought formation is still too vague. Nevertheless, it rejected the notion of thought as computer program.

Only a comprehensive fundamental theory, such as the unitary field theory of Whyte and Baranski (not to be confused with other unitary field theories), could have brightened the mental skies.

Without its insight into the functioning of our thinking organ, we would be still in the dark, as our ancestors were.

It is a known fact that a single drug can have, in addition to its primary effect, a multitude of additional side effects. This is true with any stimulation of the body processes: a single input gives rise to multiple outputs. The same does not occur in psychological processes. In them, a multitude of stimuli of various sources and kinds is followed by a single response—an idea, a feeling, or an action.

Similarly, in unitary science all psychological processes result from structural transformations of complexes of quantum structural aggregates, composed of a variety of sensory and motor components, taking place in the formative center of each hierarchy. Our actual experiences are due to the moment-to-moment variations in these structural aggregates when they are being symmetrized, and subsequently energized, by the normalizing process. Our experiential differences are related to the content of the memory organizations involved and the degree of complexity of the respective hierarchical level.

For stimuli coming from the external environment, if the formative center is located in the tropistic hierarchy, we experience a basic sensation; if the center is located in the perceptual hierarchy, we experience a percept; and if the center is located in the cognitive hierarchy, we experience an abstract concept, idea, or thought. The unity of thought rests on the synchronization of the processes of sensation, perception, and cognition.

Human thought can either be initiated by incoming environmental stimuli, external or internal, or from previously stored cognitive memory organizations directly brought to the cognitive center, where they form new structural aggregates (concepts). These memory structures, which contain cultural sequences from our social environment, can exert control over the cognitive structuring process—and so we have a *will.*

It is the hierarchical architecture of the central nervous system that allows us to control ourselves. The highest cognitive formative center is dominant in relation to the other centers below, and dominance means control.

By controlling the normalizing process (central process), the human central nervous system energizes and manipulates cognition. This is a great source of power but can also be a source of much disorder.

The origin of our feelings

While the conscious part of the mind was the one receiving more attention throughout history, the feeling part has remained the most difficult to understand. How can our emotions and feelings, the most subjective of all experiences, be based on physical processes? Again, current physics and psychology do not have a scientific explanation.

But all of us know, from tradition and personal experience, that there is some hidden link between our strong emotions and qualities of character on one hand and our heart and guts on the other. We not only experience this direct link; we also express it in our language of sorrow, courage, temperament, love, and hate. Phrases such as "I felt the pain in the bottom of my heart," "He alone had the guts to grapple with the enemy," "She went home with a heavy heart," "He lost his heart to her at once," and "I do not dislike him—I hate his guts" are commonly used. This association between some of our internal organs and our affections finds an explanation in unitary theory.

According to unitary biology, our emotions and feelings originate in contractions or changes in the tonicity of smooth muscle cells present in visceral organs, walls of small blood vessels, the heart, and elsewhere. These contractions are triggered by the autonomic nervous system, hormones, or other local agents. Special organic receptors in the smooth muscle cells convert the patterns of the different contractions, with their subtle qualities and momentary variations, into pulses or sequences of quantum field particles that form the internal stimuli.

These internal motor stimuli are structured into quantum structural aggregates in a special formative center before entering the ascending sensory pathway to the tropistic center. As in the case of the external sensory stimuli, these motor quantum structural aggregates decouple from the memories as they cross previously stored aggregates of similar configuration and bring them to the appropriate formative center—tropistic, perceptual, or cognitive. In the tropistic center they yield our primary drives (*e.g.*, the sensation of hunger), in the perceptual center our basic emotions (*e.g.*, fear, anger, or pleasure), and in the cognitive center our more complex and refined affects, which we call *feelings*.

There is thus a constant circle of activity from the smooth muscle cells to the formative centers of the brainstem and from these centers to the motor memory areas, the motor control centers of the hypothalamus, and, through the neuroautonomic (mostly sympathetic) and neurohormonal systems, back to the smooth muscle cells.

Conscious and unconscious states

Until the advent of unitary theory, no comprehensive view of the thinking process was available. Classical cognitive science, not knowing how knowledge is acquired and how thoughts come about, considers perception and cognition to be of a different character: perception is taken to be spatial, but cognition does not appear to be so. Because our conscious states cannot be observed, they are not apprehended as being spatial.

Despite tremendous efforts, probabilistic artificial intelligence has also been unable to reverse engineer the human mind. The algorithm presumably used by cognitive processes still remains to be found. The human brain, then, does not appear to be a natural computer that has evolved to solve computational problems.

In the unitary interpretation, what made possible the development of all mental processes was the existence of the continuous, one-way developmental process that pervades nature. Within the brain, this natural process was differentiated into three classes of unique formative processes, spatially distributed into three major hierarchies of increasing complexity, matched by a correspondent increasing order. Seen this way, sensation, perception, and cognition are not different phenomena but just particular expressions of this single natural process.

The psychological hierarchies were built from within, in a bottom-up fashion, over the course of evolution. Their processes are structural, occurring in three-dimensional space and in temporal succession. All psychological phenomena are thus resultants of spatiotemporal transactions of quantum field structures.

It took over five billion years for the cognitive architecture of present humans to be achieved: it comprises three complexing memory levels. Man, however, owes his distinctive intellectual capabilities, notably the power of abstract thinking, to the presence of the last and highest developed level. The third cognitive memory level is a unique feature of the human species.

Presumably, each composite memory pattern stored in the first complexing level contains the complete record of specific *past experiences.* This record is composed of long sequences of quantum structural organizations carrying all affective and motor components associated with each specific past experience at the time it was recorded; each stored memory organization in the second complexing level contains very long sequences of similar quantum structures that have undergone generalization. These generalized past experiences are called *concepts;* and each of the very complex quantum memory organizations stored within the third complexing level contains concepts involving language symbols. Because these concepts contain the maximum of generality and representation within a very small and highly complex structure, they are called *abstract concepts.* These include attitudes, beliefs, aspirations, etc., known as *personality attributes.* On all levels, the specific sequences of structural aggregates are interrelated via configurational structural similarities.

The ascending stimulus pathway from the perceptual formative center on its way to the cognitive center may cross one or all cognitive memory levels. On crossing these levels, the stimuli decouple similar configurational patterns and carry them to the cognitive center. It is in this center, at the moment of symmetrization and normalization, that the above concepts are experienced. So we experience percepts at the perceptual level and concepts at the cognitive level. Both result from transformations of quantum structural aggregates at different hierarchical levels of complexity.

The thinking process involves mainly the first two cognitive memory areas and the cognitive structuring center, all interconnected by the central process, which is the feedback pathway coming from the cognitive formative center and going back to it. These structures form a thinking circuit in continuous operation. When operating above threshold, it forms the basis of the human symbolic-conceptual thinking. It is from transformations of quantum structural aggregates carried from the memory levels to the cognitive center that all our learning processes, creative thinking (imagination), behavioral motivation, and all attributes of the personality derive.

The thinking circuit above described is interrupted by the *attention process,* which is evoked by normalizing distortion brought about either by external stimuli or by internal affective change. Under these circumstances, the central process extends downward to interconnect all hierarchical levels: perception and cognition work now closely together. It is during the attention process that we reason, judge, make plans, anticipate stimuli, intend to make movements, etc.

Apparently, the quantum structural organizations contained in the first memory level, the lowest cognitive level of complexing, undergo elaboration only when the attention process is evoked or when a person is asleep or dreaming.

Much has been written about the elusive but most revered mental activity known as *the unconscious*. Although the idea of unconscious thoughts—thoughts outside conscious awareness—dates back to antiquity, it was only in the seventeenth century that the first scientific explanation of the unconscious mind appeared. Being somehow connected to the brain structure, the unconscious mind was a challenge to the old body-mind dualism. In effect, its conception opened the way to the later development of unitary process thinking.

From the various notions of unconscious thought, the best known is probably that developed by Sigmund Freud (1856–1939), which formed the basis of his psychoanalysis. This original thinker considered the unconscious to be a storehouse of actively repressed instinctual desires and needs, directing the thoughts and feelings of the individual while operating well below the perceptual conscious mind. In this theory, conscious and unconscious mind were separate (and, in a way, antagonistic) entities without any coordinating principle of development. In this sense, Freud's thinking was still dualistic and static, at a time when process thought was already in development. But in the absence of a clear understanding of a formative process, it would be extremely difficult for this worker to develop a new theory of the human mind without basing it on permanent elements that could be identified.

The glory of the unitary description of unconscious processes was to belong to Leo Baranski, a genial American physicist and psychologist of the last century. In unitary psychological science, the unconscious is no more than the thinking circuit working below threshold. It is a continual organization and ordering process that underlies creative thinking. I consider Baranski's theory of unconscious process to be one of the greatest advances in modern psychology.

If you have a specific problem to be solved or need some cloudy event clarified, leave it to the unconscious. You do not have to do anything else but bear the resolvable questions in mind. The incubating period may take anywhere from days to years to complete, but a solution or a more logical explanation will always be found. Obviously, critical evaluation of all unconscious insights must be performed before accepting them.

The unconscious process is a great source of power at our disposal. Many of us use it routinely in an automatic way, but everyone should know about this hidden process specifically and be encouraged to use it frequently, particularly the young.

How about that powerful but all-mysterious entity called *consciousness?* What is the unitary interpretation of it? In unitary science, consciousness as an entity does not exist. Our awareness is due to a high degree of tendency toward symmetry in the structured field. In other words, high concentration of structural asymmetry and a strong tendency to move it toward symmetry yield the phenomenon called consciousness. The stream of consciousness derives from the continuous and rapid rate of coupling of the structural aggregates of the various modalities onto the asymmetry chain. Since several degrees of awareness may occur, there is no single condition called human consciousness.

Consciousness is a general property of the structured field and, as such, not unique to the human mind. Wherever structural asymmetry exists in the universe, the phenomenon of consciousness may also exist in some degree.

Balance of organism-environment

An organism is set in its environment. The environment continuously stimulates the organism, and the healthy organism reacts to this stimulation by progressively conforming to the conditions of its environment. This cyclic sequence constitutes the adaptive, or learning, process. Perception and cognition are just two phases of this continuous process of adaptation.

A key element in the adaptive process is the feedback pathway of memory formation, which continually updates the recorded quantum field sequences, progressively conforming them to incoming environmental stimuli.

In the case of living organisms, the environment may be external or internal; the external side is where physical and sociocultural processes operate, and the internal side is where organic processes are at work. These environments are connected to the central nervous system by sensory pathways.

We have previously discussed the sensory pathways emanating from sensors located in smooth muscle cells that transmit to the central nervous systems patterns of smooth muscle contractions that are ultimately perceived as emotions and feelings. Similar sensory afferents also emanate from myriad tiny sensors, collectively known as *proprioceptors,* located in skeletal muscle cells and associated tendons, ligaments, and joints. These sensors, or receptors, are very sensitive to small changes of pressure and tension induced by skeletal muscle contractions. It is through the patterns of these contractions, emitted in the form of pulses of quantum field particles, that we sense the size, shape, and weight of objects. It is also through them that the muscular coordination required for purposeful movements is achieved.

On their way to the tropistic formative center, the afferent pathway emanating from the proprioceptors and the afferent pathway emanating from the smooth muscle sensors are combined in the motor formative center previously described. Here they are structurally transformed into quantum structural aggregates—the motor components of internal stimuli—before they join the ascending sensory pathway of external stimuli.

In the periphery of the organism, the efferent motor pathways comprise two distinct divisions: the *somatic nervous system,* which innervates skeletal muscles and tendons, and the *autonomic nervous system,* which is connected to the endocrine system and innervates smooth and cardiac muscles and glands. These external divisions are intimately connected, functionally and anatomically, within the central nervous system.

In the unitary interpretation, thinking organisms represent the highest organizational level of differentiation achieved by the normalizing process on our planet. It is the normalizing process that holds all the differentiated systems together in cooperative balance, each differentiated system representing a component of the total process.

The comprehensive normalizing process, emanating from the dominant cognitive center, is the ultimate controller of all organic processes. Under this view, the somatic and autonomic systems, hormones, enzymes, coenzymes, and ATP are all agents of the normalizing process, operating at the different levels of the organic hierarchy—from the psychological, through the physiological and biochemical, to the field level. In other words, in a real organism the processes of physics, psychology, physiology, and biochemistry function as a single whole.

Through the somatic nervous system, the normalizing process eliminates deviations from the norm, which arise mainly in the relation of the organism to the external environment and may be said to control the *external balance.* Through the autonomic-endocrine system, it eliminates internal deviations from the norm and in this way controls the *internal balance;* that is, the steady state of the fluid matrix. These two systems are in intimate interaction, and their balanced coordination is sustained by the comprehensive normalizing process of which they are components.

To get a glimpse of the way this interaction is achieved, we can consider the regulation of body temperature as an example. If ambient temperature increases and core body temperature rises above the set value, first the blood flow to the skin increases and with it the transport of heat from the core to the skin; secondly, there is a greater secretion of sweat, which cools the skin and creates the temperature gradient necessary for heat to be lost. Inversely, if body temperature drops below the normal value, the loss of heat is cut back, and heat production is raised by voluntary muscular activity and by the shivering process. In this case, the sweat glands and the shivering reaction are structural-functional organizations that the normalizing process has developed to offset normalizing distortion due to extremes of temperature.

A steady composition of the fluid matrix is critical for the operation of the cellular enzymes, including those of metabolism. Through the secretion of hormones and their control over enzymatic action, the normalizing process maintains a constant supply of structural asymmetry, which ultimately finds its way to the ATP system. This permits the normalizing process to drive synthesis on all levels of the organizational hierarchy, thus maintaining the living process in operation.

It is through the somatic and skeletal muscle systems that locomotion and manipulation are possible and that some autonomy has been achieved by higher organisms. In the case of humans, manipulation has also allowed the use of tools and scientific instruments that have led to the progressive conquest of the external environment.

Frequent repetition of the same movements leads to progressive adjustments of motor memory organizations and increased muscular coordination in motor centers. It is in this way that skills are learned and then performed in an automatic way, as seen in sports and in the performing arts. There is, however, an acquired skill that stands out as very special—that of spoken language.

4

THE PROCESSES OF LANGUAGE

If there is a single characteristic that makes humans unique among all animals, language is certainly the one. Humans are capable not only of preserving and organizing experience but also of communicating it to other humans.

Morphologically, it was the genetic acquisition by the human species of the third cognitive memory area that allowed the development of language. Presumably, language stimuli are structural configurations of great complexity that can only be recorded in highly differentiated memory structures. Animals below man can think and reason much like humans do, but their thinking process does not involve language concepts, at least not in any great degree.

Language was developed primarily for communicating the direct experiences of everyday living. Our distant forebears first expressed their intuitive awareness in nonverbal form as body language. This basic language was, and still is, much more powerful than the enchanting verbal language, because it conveys a greater range of experience. Much of what is felt and perceived cannot be expressed in words.

Verbal language is a complicated but flexible system of noises that gradually evolved from primitive cries by separating them from their context until they became symbols for particular situations. Written language, however, had to be invented. Today, there are close to seven thousand languages of different structures spoken on our planet.

But a language is not simply sounds and spellings. It is also a whole range of affective and cognitive individual reactions to those sounds and spellings. And these reactions are different from person to person, even for those speaking the same language.

Furthermore, language is too restricted in its degrees of freedom. It must proceed in one dimension, from word to word, and thus cannot present simultaneously all the transformations occurring in process.

Despite its drawbacks, language has become an indispensable tool of human life. It has made possible the accumulation of the past experiences of our own species that have greatly facilitated and enriched our lives. And it has promoted the cooperation and coordination of efforts that are necessary for the functioning of our societies.

It is through language that we ordinarily express our thoughts and feelings. But words can sometimes hurt people, even without intention of doing so. It is therefore of extreme importance to know first how to think and then how to use the appropriate words to express our thinking clearly.

The power and limitations of symbols

Up to about thirty thousand years ago, before language was developed, our ancestors lived in the real world of things. They were intellectually undifferentiated from nature. Communication was carried out by example and mimicry. Measurements were possible only by direct comparisons between things. For instance, the primitive shepherd kept track of his flock by making each sheep correspond to a knot made on a cord. The number of retained concepts was small, memory short, and intellectual development slow.

But with the gradual rise of language a new world emerged—the world of symbols. Today we live in two interacting worlds: the world of words and numbers and the real world of things. And because of the flexibility and easy manipulation of symbols (they are spatial sequences of quantum structural aggregates), the symbolic world has become by far the largest. We can no longer think without words. They are major components of our thoughts.

For the human race, the advantages of acquiring a symbolic cognitive process were enormous. It was the equivalent of adding an extra dimension to human thinking. The spoken word made person-to-person communication easier, promoted cooperation among humans, and greatly accelerated the development of agriculture and stockbreeding. Human beings no longer had to find food for their survival.

With interpersonal relationships increasingly facilitated, the spoken word quickly grew in power. This fact, combined with the assurance of an abundant food supply, gradually led to urban living and later to the development of the great empires.

The invention of script further accelerated intellectual development. Man became capable of more general ideas; he turned introverted and was more aware of himself as a person, thus separating more and more from nature.

With the use of symbols, laws and ethical systems were established to control human behavior and maintain order in society. And with more leisure time available, philosophy, science, and mathematics on one hand, and literature and the arts on the other, made their appearance.

The discovery of the quantitative scientific method around AD 1600 was followed by the technological revolution and the exponential growth of all the sciences and arts. And over the centuries, each generation has been passing the accumulated knowledge to the next. Civilization as we know it would not have been possible without a symbolic process of thought. However, we should never forget that all our mental processes were ultimately developed, organized, and guided by the formative-normative process of the field. The order embedded in our symbolic world is a continuation of the order of nature: the order of the physical events in the real world.

But in spite of all their power, symbols such as words also have their drawbacks. One of those is their *incompleteness:* they cannot describe fully either the real things they stand for or the related cognitive processes. On the other hand, words can carry emotional content and can be manipulated in ways that things cannot.

Words are abstractions or essences of things and, as such, leave out most of their characteristics. In the abstraction process, similarities are captured and differences ignored. The word "chair," for instance, denotes all objects having a seat, a back, and legs. To differentiate and describe a specific chair, more words, and more thinking time, are needed, and the description, in all its details, will never be complete. Without the process of abstraction as the first step, no further acquisition of knowledge is possible.

Regarding living things in process—say, an apple tree or a dog—the situation is more complex. The static words can only capture their changing patterns at a given instance in time. Today's "dog" is not the same "dog" that we will find the next day. He will have grown in size and become more skilled and knowledgeable. Static words are not appropriate for describing ongoing processes. They confer a sense of permanence and stability to constantly changing events.

Words never say all about anything. This is the reason we keep writing books about the same subject over and over again. The limitations of words are quite evident in the description of our affections. The infinite differences in our feelings toward all the many experiences we undergo in life are too subtle to be described by words alone. So we express them by complicated manipulations of tone, volume, and rhythm of voice and other affective devices of language.

Because words are abstractions of one's concepts, they have individuality. And since our memory contents are necessarily different from person to person, on account of different genetic and cultural backgrounds, past experiences, likes and dislikes, degree of cultural development, etc., the meaning of a given word is not exactly the same for each person. The meaning also changes according to the context, which is the entire surrounding set of circumstances under which words are used. Most words have more meanings than those described in the dictionary. In effect, as Samuel Hayakawa, a renowned teacher and semanticist, emphasized, "No word ever has exactly the same meaning twice."

Since it is through words that we must communicate, alterations in meaning may create problems in daily conversation. The person we talk to may become offended by something we say, although no offense was ever intended. It is the subjective meaning of words, not their sounds, that people react to.

Ambiguities derived from different meanings complicate the understanding of scientific concepts, resulting in much confusion if one is not aware of the specific differences in meaning. Psychologically, the meaningful aspects of our symbolic-conceptual ideas, attitudes, assumptions, values, etc. are due to interrelations of sequences of underlying memory organizations. Changes in meaning must be met by reordering of these interrelations, a process that may take some time to be accomplished.

Much more time—in effect, years of learning—is required to achieve fluency in reading, a process whereby we rapidly translate print to meaning. When well developed, the reading system takes only about 0.05 seconds to identify a word from the 50,000–100,000 we generally know.

Languages are enormous systems of classification made possible by the abstracting and generalizing capabilities of human nervous systems. The generalization of the symbolic concepts results from the continuous structural development of the cognitive memory organizations during thinking. The end result of this ongoing process is ever-finer differentiations of behavior, expressed in motions, language, and culture.

Because of the extreme complexity of our social and semantic environments, we need to develop the habit of delaying our responses, allowing more time for our symbolic thinking activity to recognize similarities as well as differences in the world around us. This way of thinking was in the mind of President Thomas Jefferson when in 1811 he wrote a letter to his twelve-year-old granddaughter advising her on how to travel unscathed through life. The letter contained twelve tips, the last one reading, "When angry, count ten before you speak; if very angry, a hundred."

When the above advice is followed, we are better able to distinguish words from what they stand for and to formulate our spoken or written responses in an appropriate and polite way.

The reification process

Besides incompleteness, words manifest another distinctive property in relation to the things they symbolize: *independence.* There is thus no necessary connection between words and what words stand for. An abstraction in our mind may not correspond to, and may be confused with, a thing in the real world. We can, for instance, assign a name to a nonexistent thing and act as if the abstracted name were the thing in the outside world. This process, the tendency to objectify our abstractions, is called *reification.* We tend to think of terms such as power, justice, social class, interest group, democracy, and so on as referring to real things in the real world.

The reification process started very early in antiquity during the development of human thought and may have originated from still another characteristic of words: they impart a sense of constancy and stability to continuously changing events. This attribute led Plato to believe that purity and beauty could only be achieved in the abstract world of words and not in the chaotic world of things. For him, the persistent harmony of the ideal world was reality, and the confusion of the world of process was illusion. True knowledge could only be achieved at the top of the abstraction ladder. In other words, the farther he was from nature, the nearer he thought to be to the truth.

It was from this thought process that the ideals of beauty, truth, perfection, goodness, and others derived. But there is no beauty without a "beautiful" thing, and what appears to be beautiful to someone may not be so to someone else. Similarly, the admission of perfection is just the denial of the possibility of further growth. All ideals are no more than attempts to interpret the development process as a group of tendencies permanently directed toward certain unchanging and universal ends.

But the development process, as understood by unitary theory, is not directed toward any specific end condition. There is no certainty in life. Yet the idealist desires to escape the uncertainty of process and to find spiritual security in the aim of a permanent harmony. He fails to realize that development, not permanence, is the factor that stabilizes and harmonizes all the component processes in the individual.

We have previously noted that what the human brain perceives in the outside world is structure or form. The stimuli emanating from a structure undergo a series of transformations in the brain, ultimately yielding ideas or concepts and the words expressing them. Ideas are thus functions of structure and not the other way round. Saying the same thing in a more fundamental way, the quantum structures of the field are the empirical referents of every concept, since only they are endowed with causal properties. In the reification process, the causal relation—the relationship between the empirical referents in our environment and the correspondent symbolic concepts—is faulty. In the above example of the Platonic ideals, the empirical referent is displaced from the environment to the mind—the causal relation is reversed.

Basically, the special characteristics of human thought make reified concepts possible. Besides its capacity for abstracting and generalizing the data of experience, our thinking process can manufacture symbols at will. The normalizing process can decouple and bring recorded memory sequences, or parts of them, directly to the cognitive center without the involvement of any external stimuli. The structural aggregates of the concepts that are formed may or may not possess any relation to the environment.

Reified concepts do not correspond to what is "out there" in the real world—they are unveridical. These concepts, which can carry emotions and feelings, exist only in the mind of the person possessing them—they are cognitive illusions. And to complicate matters even more, the person is seldom aware of their existence.

Because of their unveridical nature, reified concepts are distortion-producing. With time they tend to block the learning process, thus preventing their selective elimination by the normalizing process. The formation of new memory sequences that could facilitate development in relation to the environment is also barred. The affected person becomes progressively isolated from his environment, and change, learning, and healthy development cannot take place. Reified concepts are thus very debilitating. No wonder reification is a potent source of confusion, irrationality, and arbitrariness.

Generally speaking, there are two major ways of acquiring unveridical symbolic concepts: from our own brains (our thinking and reasoning) and from other people's brains (our social and cultural environment). In the latter case, we get them from friends, educators, newspapers, TV, radio, movies, books, etc. We are literally surrounded by reified concepts and must be aware of their existence if we are to keep our sanity intact.

In daily life, the process of reification may begin in misunderstandings of some simple facts of the environment, in wrong and usually unconscious assumptions, or in words or events taken out of context. Limited knowledge of a particular phenomenon or issue may be an underlying factor. Occasionally, the reification is due to continued use of primitive and mistaken expressions, such as those of "sunset" and "sunrise."

Included in reified concepts are those abstracted words carrying special connotations, usually qualities, that are wrongly attributed to specific persons by generalization. We tend to generalize people by similarities while ignoring differences. These are the cases of so-called preconceptions, or *prejudices.* A most illustrative example, given by Hayakawa, is that of the word "convict". Most people assume, wrongly, that all convicts are criminals but some may later prove to be innocent of the charges brought against them. Similar generalizations are responsible, at least in part, for many of the attributions in society regarding race and nationality. But there are no absolute identities in the world—person A is not person B.

Everyone is different in several respects, and therefore a single label does not apply to all.

Another example of overgeneralization is that involved in commercial propaganda. The product advertised is the best there is, and no others are even comparable. Rather than limiting itself to giving useful information about the specific product, modern advertising associates the product with all sorts of magic connotations. Then the advertisement is continuously pounded on us. On the TV screen and on the radio, advertisements are thrown in systematically every few minutes, in drop-like fashion, no matter how important the background issues being presented are, to our lives or to the future of the country. The pounding is intended to give us automatic reactions, to force us to say yes to brand names invested with all kinds of wonder, delight, and thrill; it is deliberately tailored toward blocking our judgmental and interpretative processes, and directly influencing our patterns of evaluation. Independent of the quality of the advertised products, the timing of present-day advertising is outright insane, let alone its content.

But reification of concepts is not limited to colloquial language. It is embedded in the cognitive systems of religion and philosophy, and even in those of present science. The empirical referents of religion are found in the holy books, and those of philosophy in logic and reason. Both of these disciplines are thus devoid of denotable empirical referents. So are present physical laws. They were made by physicists, based on measurements of certain quantities in the real world. They were not directly derived from nature—the causal relation is also reversed here. Unaware of the natural process, classical physicists unknowingly applied the Platonic approach to the environment. Although no constant of nature has ever been derived directly, further developments of unitary theory should allow us to do so.

Looking back, over two millennia of total time, and half a millennium of scientific time, have passed since Plato walked on this planet, and reified concepts are still with us today. They have pervaded all levels of knowledge and are now everywhere—in our individual, social, and cultural lives. As a result, they constitute potent sources of general disorder.

We are now more acutely aware of conceptual-linguistic reification than ever before, because unitary theory had the power to expose it. Using unitary concepts is like taking a look at the earth from outer space: things are viewed from a broader perspective. The time is ripe for us to take off the straitjacket of irrationality that has prevented the human species from direct connection to nature. For this to occur, we must accept process without reservations. Only then can healthy intellectual development take place.

But the task facing us is not going to be an easy one. Reified concepts are now spread over our memory organizations, motivating the way we think and behave. A lot of unlearning has to be done before the new concepts can be assimilated. We may start by giving up the old unveridical and wholly quantitative cognitive systems and adopting the new qualitative, scientifically based, veridical cognitive system.

This does not mean that we should always abstain from forming concepts unrelated to the environment. This would be to negate the functions of our creative processes that have given us the arts, the mathematical sciences, and all of our imaginative life. It applies only to circumstances where the symbols must represent the empirical referents. In these cases, the symbolic concepts must sustain, or progressively achieve, a one-to-one correspondence to those empirical referents.

It is somewhat ironic that language, which has allowed the human species to reach such a high level of dominance in the animal kingdom and given us the means of communication that led to the present level of civilization, has brought us a lot of misery too. As Goethe hinted centuries ago, "There is something foppish about it." It is this "foppishness" of language that we must understand and correct.

The language of agreements

Humans live in a complex world of words—the semantic environment they have created—and must depend on language for communication. And with civilization advancing, the need for communication is even greater. It is therefore vital that all of us agree on what the words we use stand for. Without agreement there can be no cooperation, and without cooperation no progress is possible. If language were to become meaningless, civilization would be brought to a halt.

But given the complexities of our semantic environment, how can agreements be achieved with words? Is there a technique that can guide us to agree on something? And do agreements on what words stand for always guarantee the veridicality of the underlying concepts? These are not easy questions to answer.

The need for some order in language was recognized early in the history of civilization. To this purpose, Aristotle developed a set of rules—including those of two-valued, yes-or-no logic—for governing consistency in the use of language. He arrived at these rules by the process of abstraction, and therefore Aristotelian logic is language about language and not language about things or events. It lacks the principle of denotation, and as a consequence, the veridicality of logically deduced statements is not assured. Hayakawa states this conclusion more vividly: "They may be 'accurate' maps of real 'territories' or they may not." As time went on, the Aristotelian logic was improved by other philosophers, but because of its very nature, it never led to agreement among them.

One way to agree on what words or other symbols stand for is to establish these agreements beforehand. This was the approach taken by arithmetic and geometry. The logic of mathematics is the Aristotelian two-way logic of "true" or "false," with preexisting, clear-cut linguistic agreements as to what the X, Y, and other symbols stand for. To be sure, these symbols are all abstractions, but they are properly and uniformly made from starting points in the real world. For this reason, mathematical symbols are useful in predicting occurrences and in getting work done.

Mathematical logic cannot be used in ordinary conversation where only the vaguest of agreements exist between the speaker and the recipient as to what words stand for. Moreover, two-valued logic is too restrictive: it allows only two choices—yes or no, black or white—for a single decision. Everything must be accounted for under one of the two values. There is no other way of evaluating new experiences and of increasing one's knowledge. The rigidity of this primitive type of logic, tailored for an emotional and superstitious life, is not appropriate for modern humans with well-differentiated cognitive systems, harboring a gradation of many interests and desires, each requiring a specific way of acting. It is thus not surprising that a conversation based on two-choice-only logic inevitably leads, sooner than later, to disagreement and ultimately aggressive behavior.

A more flexible, multi-valued logic, encompassing a range of choices, is more appropriate for ordinary conversation and is the one used in intelligent public discussions and in the political discourse in democratic societies.

In the beginning of the eleventh century, during the Middle Ages, Ibn al-Haytham, a renown Muslim scientist, worked out the basic requirements for a truly scientific method of investigation. But only around 1600, when its quantitative basis was finally established, did the present scientific method become fully operational.

To satisfy the requirements for high precision in measurements, science adopted an infinite-valued logic based on fixed and previously agreed-upon scales of values. Temperature, for instance, is given not in hot or cold but in degrees, and speed not in fast or slow but in kilometers per hour or meters per second. With this type of logic, one's actions can be precisely adjusted in an infinite number of ways. And with fixed frames of reference, general agreements are possible, as in the case of mathematics. As a result, science has developed quickly and has led to a multitude of inventions.

But the scientific method does not comprise only machines and standardized measurements. These are just the means whereby correspondence between words and things is progressively achieved. It contains within itself a certain discipline of thought, a specific orientation toward the exploration of reality. It involves observing, making hypotheses, and testing them. The results obtained are checked by a number of independent observers, rechecked again, and corrected if necessary, before general acceptance. And no results are ever final. They can be challenged at any time, and science welcomes the challenge. This type of humility, a product of emotional maturity and intellectual flexibility, underlies the determination of science to finding bases for universal agreement. No other body of knowledge possesses this kind of orientation.

Given the reliability inherent in the establishment of veridical results, it is recommended that only concepts obtained by the scientific method be trusted and accepted by the men and women of the twenty-first century.

But the present scientific method of investigation is not the only one possible. It was just the first heuristic method to be developed and institutionalized. Unitary theory is proposing a second heuristic method, a new approach to measurement that uses a relational function, still to be developed, based on order rather than quantity. The unitary concepts are highly challenging to classical science, but given the conceptual openness of its orientation, scientists will sooner or later recognize their power and accept them for proper testing and further development.

This would be very fortunate, since the present quantitative method of measurement, after four centuries of exploitation, has very much exhausted its capacity for obtaining new knowledge—most of what needed quantitation has already been quantitated. This is particularly true for biology, psychology, sociology, and medicine, the so-called natural sciences. Furthermore, as a result of the use of ever more potent telescopes, microscopes, and automatic technologies, a tremendous amount of knowledge has been accumulating at an increasing rate, almost exponential over the last half century. Since the quantitative method is one of analysis and not synthesis, all this knowledge remains loosely dispersed and disorganized, waiting to be put together in some orderly way. This chaotic situation is slowing down progress, with adverse effects on all humanity.

The more powerful method of investigation proposed by unitary theory would solve these problems. It would not only extract new and perhaps more relevant information from structure and study the particular forms and development of systems, but it would also integrate the obtained and already available knowledge into a comprehensive whole. Human intellectual development would be greatly accelerated and a higher understanding of reality achieved. And all of this should bring humanity closer together.

The communication process

Humans are primarily cognitive beings with very gregarious habits, high manual dexterity, and muscular strength strong enough to break molecular bonds in solid materials. These characteristics enabled them, early in evolutionary history, to make tools and use them for gathering food and building shelter. The skills acquired in these activities were essential for living and were passed on by the spoken word, together with rituals and myths, from one generation to another for thousands of years. For humans, communication was, and continues to be, the central mechanism of survival.

With further intellectual development and the spoken language recorded in script, communication was greatly accelerated and soon led to a progressive accumulation of a conceptually organized social tradition. This development changed the characteristics of the communicative process.

Before the establishment of a community tradition, the recipients of knowledge were the tribal elderly, and the communication process was one-way, from ancestors to descendants. Ancestor worship in primitive societies was a way of assuring, recognizing, and respecting such communication. With a functioning tradition, communication became an interactive two-way process.

The individual enriches the community tradition with new ideas and forms of behavior, and the tradition molds and matures the individual, enabling him to carry the process further. But whereas the individual has a short life span, the tradition long retains the records it has absorbed from individuals. The social tradition functions as a progressively changing environment that brings out further capacities in the young of each successive generation.

When we make use of science, technologies, law, and literature, we are using the accumulated observations, insights, and discoveries of thousands of fellow humans, most of them long since dead. Communication from past to present binds all humans throughout time.

The individual-tradition-individual cycle of the communication process has been maintained for at least the last five thousand years and is responsible for the fast intellectual development of humans, and all the civilizations they have created, over this span of time. Periods of stagnation, and even regression, have occurred, but this general trend has been maintained and, barring a globally destructive natural or human-induced catastrophe, is expected to continue in the future.

But we should never forget that what makes communication possible is the quantum field process working in our psychological memories' organizations and formative centers—the force behind existence.

5

THE SOCIOCULTURAL PROCESS

It was the need for self-defense and survival that forced our distant ancestors to associate in small groups—the primitive tribes. Over time and with improvement of life conditions, increase in knowledge, and development of new technologies, the social groups proliferated greatly, and some of them differentiated in the performance of specific tasks. Modern societies are made up of a multiplicity of social groups, all interrelated and interdependent in their functions. It can be said that the social group, the aggregation of humans for common purposes or shared goals, is the basic unit of every society.

There are similarities between a social group and its counterpart in the physical world, the physical system: the parts that form the whole are not free to move about. They are constrained in their movements (tendencies) by dominant relations of whole over parts. But whereas the physical world deals with space-time relations, the social world deals with behavioral relations.

The social relations have the same significance and are as real as the physical relations. But the latter take place in a circumscribed space and are time-invariant, while the former are separated in space and may change with time. Both are relations, nevertheless.

To be effective, relationships must be based on some kind of order; but humans, unlike physical parts that always obey the natural order, are thinking organisms and, as such, free to make decisions, which may or may not obey the required order. Social order therefore must be enforced. In primitive human societies, order was maintained by physical coercion, at the mercy of the lasher, but soon man learned that words—written codes of law—could do the same and in less arbitrary ways.

Besides that imposing legal order, there is something else in a full society, such as the modern one we live in, where many specialized social groups are interacting and voluntarily cooperating with each other. Some kind of mutual, unspoken agreements bind all societal groups together. For instance, some group agrees to deliver a specified product, and we agree to buy it or pay for using it; we agree to observe the rules of a club, and the club agrees to let us enjoy its privileges. With these agreements, behavior is disciplined, cooperation is encouraged, and peace is achieved; without them, no civilized society would hold together.

The above agreements are, or should be, based on truthfulness, honesty, trust, and loyalty—qualities that are part of the so-called human values. These values, or norms of conduct, were developed over the centuries, through the molding of our feelings. They have been accepted by science and are now taught to children by parents, teachers, and the institutions of society. They are the foodstuffs of civilized human personalities.

But how did human values actually arise? Where do they come from, in the first place? They must derive from special memory organizations established by the normalizing process, and used in activities promoting human unity. The order contained in them is ultimately derived from the universal order.

Societies were developed by individuals, and individuals are guided, whether they believe it or not, by the ubiquitous universal process. But this process is still unknown to many sociologists, and this fact is responsible for the different views of sociocultural development.

Factors in sociocultural development

Modern non-unitary theories of social development, such as systems theory, are still based on Darwin's ideas on the struggle for survival. Development is thought to result from competition between societies or social groups. Competition tends to select the most efficient societies, which are also the most adaptive ones.

With increasing division of labor resulting in thousands of different occupations, social complexity increases greatly. The selection pressure forces societies to adapt to this situation by increasing their functional efficiency. If this adaptation fails to occur, societies will tend to be eliminated over time.

The overall tendency is toward greater complexity and efficiency, although perfect efficiency is never reached. But increases in efficiency require increases in internal coordination and communication. These requirements must be met for societies to continue growing and differentiating.

This mechanistic view of social development neglects the direct connection of nature to man, and the extension of this connection from man to society, that is at the basis of the unitary conception. It is the universal formative process (or, more simply, the normalizing process) that is responsible for the development of all forms in nature, including societies. Social structures were formed by the normalizing process working in individual organisms and using their memory organizations as immediate developmental agents. Social development is therefore an inner natural process, always working toward order and organization, and not an outer mechanistic process due to external competition and subject to trial-and-error uncertainties.

In the unitary view, social development does not necessarily require competition between societies or between social groups; it only requires facilitated conditions for it to pursue. All social organizations must be oriented toward facilitating the development of the individuals comprising a society, and the individual must facilitate the development of all social organizations of which he is a part. The individual and his social environment are constantly adapting to each other, but given their extreme and changing complexity, complete conformity is never reached. Organic development is a process of continuous adjustment. If a social organization does not adjust, it will pass away, and if the individual does not adjust, he will be selectively eliminated.

The unitary conception emphasizes general development because only through the pursuit of general development can the human species acquire a unity of purpose capable of bringing it closer together.

A glance over the sociocultural process through time

Since its appearance on the surface of this planet, the human species has made an astonishing advancement—from small groups of food-gathering and hunting wanderers living in dark and cold caves to the civilized life and comfortable homes of our modern cities and towns.

At the basis of this achievement certainly was the hereditary constitution of the species, which allowed it to acquire a symbolic process of thinking and, later, a language. These faculties then enabled humans to interact with each other, to develop emotional bonds, to cooperate in their struggle for survival, and to form societies.

At the beginning, the need for subsistence forced humans to aggregate in nomadic groups or tribes that shared the food collected—seeds from wild grass plants, fruits, nuts, game, and fish—on a daily basis. Speech slowly developed, but the higher brain functions were still rudimentary. The human individual was unaware of self as a being separated from nature. Incapable of rational thought to explain natural phenomena, he turned superstitious. There was no organized political power in these early societies that existed somewhere from two hundred thousand to twenty-five thousand years ago. But from the ties that bonded their members, the first social unit—the family—slowly emerged. Most members of the more recent tribes were related by birth or marriage.

With the passage of time, domestication of wild edible plants took place in several parts of the globe. Later, and in parallel with plant domestication, the domestication of wild animals also occurred. By 5,000 BC a new type of society, the agricultural or traditional society, based on farming and stockbreeding, was already fully developed.

There was now plenty of food available for everyone, and fewer people were required to produce it. As a result, societies grew in number, permanent villages were established, and some individuals became craft workers, producing tools, weapons, and jewelry. With the division of labor and with technological advances, the organization of society was becoming more complex. There was an increase in thinking ability, but intellectual development continued to be slow. Only permanent ideas, such as those related to daily survival, could be conceived. Memory was short.

The production of goods encouraged trade, but with some families acquiring more goods than others, social inequality originated. These families often gained power through their wealth, which was passed on from generation to generation. In this way chieftainships later emerged.

Further technological progress in agricultural toolmaking was followed by the cultivation of crops over larger areas, the appearance of towns, and further diversification of labor. There appeared a system of rulers and priests with high social status, together with a minority of craftsmen, educators, and merchants. But the vast majority of the population continued to be farmers or peasants.

As villages and towns expanded to neighboring areas, conflicts between communities inevitably occurred. The rulers organized warriors to protect the peasants from invasion, the warriors being paid with food supplied by the peasants. Increasingly, a unified, hierarchical social system of classes developed: the ruling class (warriors and priests), a small middle class (mostly craftsmen), and the peasant majority. Only the ruling class possessed political authority, expressed as military force and legitimated by the priesthood. The major task for traditional rulers was the maintenance of social cohesion, which was imposed vertically from above, and by way of physical coercion and ideological propaganda.

By 3,500 BC, human activity was very intense in the Tigris-Euphrates plain in the Middle East. These large rivers were capable of supplying abundant water and fertilizer for the cultivation of large extensions of crops. The human action was supported by animal labor and by the use of improved agricultural tools and efficient irrigation systems. Later, the introduction of the wheel facilitated the transportation of farm goods and the manufacture of pottery. Commerce expanded to nearby lands, particularly to emergent agricultural groups in the Indus Valley in Central Asia and in the Nile Valley in North Africa, and to other regions around the Mediterranean coast. The political structure became more organized, with firmer class divisions and greater separation between rulers and the ruled.

Meanwhile, writing began to evolve, allowing people to better organize their intellectual lives and more easily convey their value systems. Administrative affairs, such as the recording of data, became greatly facilitated, making an effective taxation system possible. By relying on record-keeping, merchants could extend the geographical area of their trading. New forms of economic and political organization were developed. On the cultural side, architecture bloomed, and literature made its appearance. All these changes, together with the need for coordination of a regional network of villages, led to the creation of *civilizations,* which were aggregations of city-states governed by warrior kings of great power. Increasingly, humans believed in their own potential and in their role in history.

Civilizations spread rapidly in Asia, North Africa, southern Europe, and in several places around the Mediterranean coast. Each made important contributions of its own, not only to agriculture but to intellectual culture and to the arts as well. One of those, however, was to achieve special importance for philosophy and science—the Greek civilization. Although a few Greek philosophers saw an orderly change in nature and viewed man as partaking in that order, the great majority, notably Plato and Aristotle, viewed man subjectively: human thought was separated from the world of phenomena. While nature was in chaotic change, man's ideas were stable. Truth could only be found in the static harmony of thought. This way of thinking was the one that survived, because among other factors, it fitted well with the dualistic, static religious idea of the separation of soul from body. So, without a well-defined notion of process available, a transcendental world of permanent ideas was created, which infiltrated the tradition and went on to influence human thought and behavior until today.

In the course of time, and as a result of invasions by nomads from other regions, the early civilizations of antiquity began declining or became parts of big *empires* that dominated the globe for the next centuries. Empires are vast extensions of land, sometimes spanning over two or three continents, populated by peoples of different ethnicity and nationality and governed by a single entity: a supreme ruler.

Among the ancient empires, one stands out for its military and economic power and highly developed political and judicial systems—the Roman Empire. The Romans professionalized their military class, which protected all citizens, including the peasants, and created a system of government called *res publica,* which was the source of inspiration for some modern republics, such as those of the United States and France. Besides government, law, and martial arts, the Roman civilization, strongly influenced by the culture of classical Greece, contributed greatly to the Western world in architecture, fine arts, technology, literature, religion, and language.

But empires rise and fall, and by the middle of the fifth century the giant Roman Empire, plagued by internal instability and attacked by various migrating peoples, broke up into two independent kingdoms, the western and eastern divisions. The subsequent decline of the western division had enormous political and cultural repercussions that spread to the rest of Europe.

With the dissolution of the Roman military establishment, the peasants were left unprotected. With no easy choices available to them, they sought security under the warrior lords who also were the owners of large extensions of land. Through time, a system of mutual rights and obligations between lords and peasants was created. In broad strokes, a lord granted possession of a portion of land (called a *fief*) to a peasant in exchange for protection and some sort of service, either military or agricultural. As time went on, the system increased in sophistication, leading to the so-called *feudal society* that dominated the Middle Ages, a period spanning from the fifth to the seventeenth century.

Besides the lords, the church leaders also held great power over the peasants who made up over 90 percent of the medieval population. The peasants, also called *serfs*, worked hard, and the more money they acquired, the more they gave to the church, under the belief that by doing so they would have a better afterlife. But while the church was accumulating vast amounts of wealth, some of the serfs were dying from starvation, overwork, and exploitation. From its dominant position over mental processes, the powerful church came to control the lords as well, thus becoming the most authoritative governing body in medieval Europe.

Along with agriculture there was also intense trade, particularly between the independent city-states of northern Italy, namely Florence and Venice, and East Asia. Over the centuries, some merchant families accumulated great wealth, which was facilitated by the relatively democratic values enjoyed by those city republics. Later, commerce was further intensified by a banking system that created credit and facilitated the movement of money. By the early fifteenth century, northern Italy was one of the most urbanized and wealthy regions of Europe.

Life in the Middle Ages was complicated by a devastating epidemic, the Black Death, due to *Yersinia pestis*, a bacterium commonly present in some populations of ground rodents, such as rats, and transmitted by the fleas infesting them. Presumably, it originated in China and spread to Europe via black rats that were regular passengers on merchant ships. The epidemic, which struck Europe between 1347 and 1353, killing twenty-five to thirty million people, had great religious, social, and economic repercussions. For one, the faith in the power and significance of the church declined, and the uncertainty of daily survival influenced people to live well in the present, as opposed to in the afterlife as promoted by the church.

This more liberal thinking led to a great transformation of society in all its aspects, which became known as the *Renaissance*. It was an intellectual movement characterized by recognition of the power of education, justice, and political freedom in individual human lives. A revival of interest in Greek and Arabic works, from natural sciences, mathematics, philosophy, and religion to literature, history, art, and music took place. The center of this cultural transformation was northern Italy, but soon it spread to the southern regions as well. By the sixteenth century, following the discovery of the printing press, it had spread to the rest of Europe, where feudalism was still well established. As the movement spread, its ideas diversified in their adaptation to local cultures.

The Renaissance saw significant changes in the way the individual thought about himself or herself as a person, how the universe was viewed, and how the relation between man and god was perceived. The human being became fully aware of himself as a conscious person. Natural phenomena were explained rationally and on empirical evidence. Truth was to be found through science and not through opinion or faith. From the works of Copernicus, Kepler, Galileo, Descartes, Bacon, and others, the scientific method of investigation, based on the quantitative principle, reached its final stage. The groundwork for the scientific discoveries that would lead to a major transformation of society was in place. There was, however, no clear notion of process. Human thinking was still dualistic. Descartes, for instance, viewed consciousness as a primary, substantive reality, separated from the body. The Platonic dualism of mind and matter was still well established.

On the political side, contrary to the anti-monarchical thinking of the Italian city-states, the rest of Europe saw the appearance of the monarchies. The new monarchs were kings who took responsibility for the welfare of all society. Backed by a competent military, paid for through taxation of the middle class, the monarchs centralized power and consolidated authority. The clergy and nobles were kept at bay.

Up until the Renaissance there was limited knowledge of the geography of our planet. The existence of the American continent, with all its indigenous civilizations, was unknown. But in the latter decades of the Middle Ages, a seemingly insignificant event, the development of the *caravel* by the Portuguese, was to change our view of the globe and all people living in it. Her light weight allowed the caravel to reach high speeds, and the sails' design permitted easy maneuverability and the ability to sail windward in oceanic voyages. On the other hand, the small size and shallow keel allowed entrance to the shallow waters of riverbeds when necessary. These characteristics made the caravel the best exploratory vessel of the time.

The introduction of this ingenious sailing ship was followed by intense exploration of the world, first by Portugal and Spain and later by other western European nation-states, including England and the present-day Netherlands. During the Renaissance, direct contact was established between the western European civilization and those in Africa, the Americas, East Asia, and Oceania. The globe could now be mapped and accurate distances between places calculated.

In the process of discovery and conquest, colonial empires were established, followed by exploitation of the natural and human resources (slaves) of the new territories. There was, however, a wide transfer of plants, animals, foods, languages, and cultures between the eastern and western hemispheres, at a scale unprecedented in history. The impact of those activities on ecology, agriculture, gastronomy, and social life in general was enormous. Trade spread throughout the continents, the main imported goods being gold, silver, sugar, and spices from the Americas and silk and porcelain from China. The European economic center shifted from the Mediterranean to the Atlantic, with the Dutch city of Antwerp taking the role of Florence. Socioeconomically, the sudden infusion of silver money into the cash-starved European economy, mostly from super-rich Spain, caused widespread inflation, which drove up the cost of living. The rise in prices brought about by the excess currency in circulation fueled the growth of the commercial middle class, the bourgeoisie, which came to influence the politics and culture of many countries.

Despite the many human errors committed, humanity was progressing from immaturity, superstition, and slavery into maturity, reason, and freedom. Humans could now think freely by themselves and not be told how to think by the church or the state. With the application of the scientific method, the world could be directly observed and hypotheses made based on these observations. It was thought that all basic scientific principles were already known and that, through scientific inquiry, man could now solve all the mysteries of the universe and find the solutions to all his problems. This, however, was not so.

The objective study of nature developed the mind further than could the introspective idealism of older times, and as a result the period from 1600 to 1700—known as the *Enlightenment*—was associated with great intellectual changes, from science and philosophy to economics and politics. There was more freedom of expression, allowing for social injustices to be uncovered, criticisms to be advanced, and sociopolitical changes to be proposed. It was widely believed that the aristocratic class should no longer be the only one to rule. A more democratic form of government, directly reflecting the will of the people, was favored.

By the late eighteenth century, hell broke loose in the fiscally mismanaged French monarchy. A political crisis between classes about equitability of taxation, on a background of a contrast between the economical difficulties of the poor and the extravagant life of the court, set the stage for the social unrest that followed. Encouraged by Enlightenment ideas and a weak monarchy, the tax-overburdened lower-class majority, instigated by the bourgeoisie, officially declared itself sovereign. From this civil disobedience sprang a radical social upheaval that involved the whole nation and became known as the *French Revolution*. The absolute monarchy that had ruled France for centuries collapsed in three years. The higher ranks of the monarchic hierarchy—king, clergy, and aristocracy—

were abruptly overthrown by new principles of equality, citizenship, and unalienable rights. Throughout Europe, the power of the monarchs declined rapidly, and a transition to more representative forms of government took place. On the other side of the Atlantic, similar ideas nourished the roots of American federalism.

The continuous application of the quantitative method to objects set in place and formed of matter stimulated the tendency to view man objectively. But the objective picture of man that could be developed at the time lacked any principle of integration or form. Man was therefore regarded as a "machine," a thing constructed from permanent parts, each being a static, independent entity. The subjective view of man, which dominated human thought for centuries, was rejected. Static subjectivity turned into static objectivity.

Meanwhile, in the period between 1700 and 1750, the progressive realization of unconscious processes by Sigmund Freud fundamentally changed the way our cognitive activity was thought to occur. Processes were taking place in the mind and modifying its function. Concomitant with, but unrelated to, Freud's discovery, the developmental ideas of Darwin and Wallace were also consistent with a transformative process going on in organic species. These discoveries had a profound impact on the philosophical and scientific thought of the time, ultimately changing its concepts from static permanence to process, or, in a more fundamental way, from symmetrical relations to asymmetrical. The admission of the existence of unconscious processes further demanded discarding of the dualism of mind and matter.

During the latter part of the eighteenth century, the persistent application of the scientific method of investigation to the study of natural phenomena had led to a series of inventions that would drastically change the quality of human life on the planet, and this change still continues today. It was driven by the utilization of new forms of energy, first mechanical and then electrical, to the benefit of mankind. Human society was profoundly transformed from one based on manual labor to one based on machine manufacturing.

It started with the invention of the steam engine, powered by coal, which was applied to the textile industry, agriculture, mining, and transportation. Steamboats were now transporting goods and people through rivers and canals, and trains were running on rails on land. The price of shipping was diminishing. And iron tools were being used to construct machine tools capable of repeated production of high-precision, and therefore interchangeable, parts. These parts could be used to assemble, not one at a time, but a series of similar machines for manufacturing in specific industries. Worn-out parts were easily replaceable, and some of the parts could be utilized in different machines. In the long run, the mechanical production of goods was reflected in lower prices and much increased variety. This marked the beginning of the industrialization of society, a process that started in England, where a period of peace, an educated middle class, and business-friendly legal and economic systems encouraged investment by risk-taking entrepreneurs. It soon spread to the rest of Europe and to the entire globe.

By the mid-nineteenth century there was a fundamental understanding of thermodynamics theory, which greatly aided the development of the chemical industry. Control theory was also available and in use in many forms of automation. But it was the electromagnetic theory of James Maxwell—the unification of light, electricity, and magnetism—that ultimately revolutionized technological development.

In the period 1850–1950, wrought iron was converted to steel, and the steam engine was replaced by the more efficient steam turbine. And now large steel ships powered by steam turbines were exchanging goods between west and east. The petroleum industry made its appearance, and the electrical industry flourished with the production of generators, transformers, and motors of various kinds. The generation of electrical power by steam turbines made the lighting of streets, factories, and homes possible, and power railways greatly improved public transportation in the crowded cities. Public health and sanitation were also improving, after the experiments of Louis Pasteur confirmed the germ theory of disease. By the turn of the nineteenth century, a useful internal combustion engine was available and used in the construction of automobiles and later, with more knowledge of aerodynamics, airplanes. This was followed by the introduction of the internal combustion turbine, which made jet propulsion of aircraft and long-distance travel a reality. And the introduction of electric tools in the automobile industry created the assembly line, with its associated mass production. At the same time, with the advent of the telegraph and later the radio, communications over long distances were facilitating trade and bringing human societies closer together.

The social, economic, and cultural impact of all these inventions was tremendous, comparable only to that brought about by the introduction of agriculture and stockbreeding. Living standards improved significantly in the newly industrialized countries, particularly in the first half of the twentieth century, when full-blown economic development occurred, as prices of agricultural and manufactured goods fell dramatically due to increased efficiency of production and decreased costs of distribution. In the two centuries following 1800, the world's average per capita income increased over tenfold in spite of a population increase of sixfold. But the increase in personal income was lopsided, favoring mostly the rich. The gulf between rich and poor actually grew wider.

Prior to 1950, the entertainment industry had made its appearance with the development of film and later, television. And new insights into mathematics and logic ultimately led to the development of the computer, followed half a century later by the Internet. The computer and the related robotics industry came to play a central role in modern manufacturing and in our daily lives as well.

Between 1950 and 1970, the available technological knowledge made possible the initiation of space exploration, first with the launching of artificial earth satellites that enabled interplanetary communication, and then with man's landing on the moon. The earth-orbiting satellites later led to great improvement in weather forecasting and communications on earth, the most recent being the global positioning system (GPS). Basic scientific knowledge was expanding in most of the sciences, particularly in physics and in the chemical and pharmaceutical industries. Technological achievements also occurred in these disciplines, among them the practical application of hemodialysis. But one of the major scientific events of this period was the development of a general theory of process, the *unitary field theory*, the subject of this work. This theory brought the subjective and objective views of man together in a single insight. Nature, organism, mind, and human society became components or expressions of the same process.

It should be noted that not all discoveries lead to social or intellectual progress. As a matter of fact, most of them do not. Thousands of technologies have been invented, but only a few are worthwhile to keep: those with wider social application. Likewise, the development of new scientific theories is not always followed by advances in our knowledge.

Before going ahead, it may be worthwhile to take a retrospective look at sociocultural development until this point in history. Although, by limits imposed by the internal order of this work, the narrative is very incomplete and refers only to western civilization, to the exclusion of all others, it may allow some analysis to be undertaken using the concepts of unitary theory.

Sociocultural development as a formative process

First of all, development of new ideas is a process that requires time and cannot be hurried, as can be seen by the development of the concept of process itself. Secondly, societies are extremely complex organizations ultimately brought about by the normalizing process of the field working through the psychological processes of humans. By analogy, a society should have its own metabolic system, a circulatory system, a nervous system, etc., as organisms do, but in a very complex and highly differentiated form. What, then, constitutes the metabolic system of society? It is the economic system. Anabolism at the biochemical level corresponds to the productive process (raw materials, labor, goods, and services) at the societal level, and catabolism is equivalent to the whole structural organization of the productive apparatus (distribution, consumption, management, capital investment, etc.). The enzymatic system of society is represented by human organizational agents. Likewise, the circulatory system at the biochemical level is analogous to the whole network of transportation systems of society, and the nervous system is made of all its sociocultural organizations—schools, universities, arts, museums, television, Internet—that develop values and promote interpersonal relationships. In a general way, societies can be looked upon as gigantic and complex organisms, or huge networks of interrelated and interdependent formative processes.

Formative processes are always in constant change, but as Whyte pointed out, "Form is manifested only when circumstances permit." In other words, only when the social environment is facilitating can form, or social-economical-cultural progress, occur. Otherwise, development may not take place. This is clearly exemplified by what occurred in northern Italy during the Renaissance period and in England at the dawn of industrialization. The sociocultural stagnation that characterized the Middle Ages, particularly during its first period, appropriately called the Dark Ages, was not due to a lack of talented people. It was a direct result of its rigid, dogmatic, fear-dominated society, in which thinking had little or no contact with reality. And here is advice, derived from unitary theory, to all political leaders: the proper policy at any time is the one that facilitates development in a given environment. A policy that works in one nation may not work well in other nations made of people with different characteristics. Societal development is the result of a constant adjustment or mutual adaptation of individual and social environment, and it should be based on cooperation. And the development should aim at getting closer and closer to veridical concepts.

In a unitary process, when form develops, it will not last forever. If form follows its natural course, there is a phase of growth and then one of stabilization, followed by a phase of decay. No form is eternal. Thus, stars and hurricanes appear and disappear, civilizations and empires rise and fall, and we live and die. The evolution of the sociocultural process described above is entirely consistent with unitary concepts.

Although we tend to divide the historical process into several periods of development, it is nevertheless a single and continuous process of transformation of the human cognitive system. Under this view, it is not proper to talk about religion, philosophy, objective science, or unitary science in isolation. They are just phases or components of the same process and can only be fully understood as such. Similarly, there is only one law of development for all organizations of the universe—from asymmetry to symmetry, from less conformance to more conformance of organism to environment.

The present stage of the sociocultural process

Throughout most of the twentieth century, the scientific and economic performance of the industrialized nations was simply remarkable. Fueled by new inventions and technologies, and supported by an ever more educated workforce, productivity increased steadily, and the living standard of each generation far exceeded that of the preceding one. As a result, a strong middle class arose.

But since about 1970, for some strange reason, the pace of significant scientific discovery slowed down, in spite of exponential increases in human and financial investigative resources. The developed countries have now somewhere between five to ten researchers per thousand workers and are spending about 1.5 to 3 percent of their gross domestic product on research and development. Judging by the number of scientific publications worldwide, the amount of scientific knowledge accumulated has skyrocketed and is now beyond what the human mind can encompass. All this activity, however, has been uncoordinated and unassociated with a correct arrangement into comprehensive generalizations. Some kind of integrative insight appears to be missing. The need may be not for more knowledge, but instead for more order in knowledge.

Most of the development in this period occurred in physics, mathematics, technology, and other sciences where abstract thinking is used. No significant advances occurred in biology, psychology, and sociology, the so-called natural sciences. Most of the progress in biology and medicine has been due to the introduction of new technologies and not to significant scientific breakthroughs in the understanding of living processes in health and disease. The most important biological accomplishment of our time, the sequence of the human genome—a three-billion-dollar, thirteen-year enterprise—still remains to fulfill the purported impact on the cure or prevention of cancer and other pathologies. The continuous appearance of new technology has given us a false sense of progress.

Is there any explanation for this state of affairs in science? Is less out there to be discovered? Are there any factors in scientific research blocking the road to progress? If so, can unitary theory be of some help?

There are indeed factors of varying nature—need for higher and wider education, too much bureaucracy, government interference, and many others—adversely affecting scientific and therefore socioeconomic development and wealth creation in our society. But in this work, we are mostly concerned with more basic causes originating in the disparity between the nature and workings of our psychological processes on one hand and the nature of present science, its investigative method, and the mechanized industry on the other, as seen by unitary theory.

By its nature, unitary science pays special attention to the whole without ignoring the parts. The formative process of the whole dominates those of the parts, which "cooperate" with the whole in its pursuit of symmetry—a single comprehensive process. Present science has been operating in an inverse way. Although there is only one science, the whole, it has been artificially divided into separate disciplines or branches, the parts, to allow our limited intellect to better grasp the subjects at hand and for didactic purposes. But science went too far in this divisive process of specialization, and now a variety of subdivisions exist for a single specialty. There are serious drawbacks to this kind of approach.

As specialization proceeds, people know more and more about less and less—a tradeoff between more microknowledge and less general understanding. This leads to professional insularity and a failure of judgment, caused by a lack of perspective and balance. The superspecialists are now doing the research work, and given the inherent interrelation and interdependence of biological structure, they are in danger of not grasping the general cause of the phenomena they are investigating and of committing fundamental errors in their interpretations. The micro is part of the macro, and the latter has been ignored. To correct this problem, the scientific education of the researcher must be broadened, or the research work must be done by an interdisciplinary team.

Deep down, scientific specialization is a by-product of the intensive use and limitations of the quantitative method of investigation, which is based on the symbols of number and quantity. But the concepts of number and quantity are just abstractions devoid of causal properties. The thinking based on these concepts bears no direct relation to experience. The quantitative method is appropriate for enumeration, analysis, specialization, and accumulation of magnitude but tends to ignore form, order, and development. The method has been very successful in giving us science and technology and in developing civilization up to this point. But it appears that the slowdown of innovation, the stagnation of physical theory, the lack of basic development of the natural sciences, and the inherent disorder in our economic system are all expressions of its inadequacy.

A complete unitary analysis of the faulty quantitative method cannot be given here, but a few points can be made. First of all, the quantitative method, based on static thought, emphasizes *precision:* precise analysis of the parts that constitute a whole. Process thought emphasizes *completeness* instead. Nature and man, for instance, are separated in static thought, but they become one in process thought. Paraphrasing Whyte, there is a change in aim from analytical precision to general order. Secondly, the quantitative method does not extract all the information from a tridimensional system. As a matter of fact, most of it, perhaps the most important, is left behind. Knowing the percentage of the different elements that comprise a system is not enough to characterize it. We must also know the whole arrangement of its atoms and atomic groups: its complete structural pattern. Finally, the old assertion that "scientific theory must advance by continually increasing abstraction" is no longer valid. In fact, under this claim, physical theory has not advanced at all. We must integrate experience with thought and action—they are interrelated aspects of the same process. We need to adopt a concept of reality that brings nature, environment, and science into a single whole.

Contrary to the mechanical processes of technology, which can be accelerated by deliberate effort, psychological processes, which underlie our intellectual and moral conceptions and are involved in our adaptation to the social environment, cannot. Mental processes require time and cannot be rushed. Adaptation to new knowledge and new technologies requires many rearrangements of our memories and new interconnections to be established among them. This is not a continuous process, and it may take a long time to be achieved. Technology has always been ahead of us, and we do not have the required time to catch up. This situation has been deteriorating steadily since the introduction of technology in our society and inevitably has led to intensification of social friction. The present-day requirement of more specific knowledge and more flexibility in knowledge of our workforce is also a direct consequence of this phenomenon. Man must adapt to the requirements of the machine. This biological-mechanical imbalance cannot continue forever.

Another disturbing consequence of technology has been the mechanization of the human being. Quantitative science, devoid of any principle of integration or form appropriate to biological and human organization, treats man as a machine. It disregards our humanness and our connection to the process of nature. The most basic of human values—individual freedom—tends to be neglected. We should strongly oppose this.

Technology had its origins in the desire for greater human autonomy, more leisure time, and better control of the human over his environment. But by placing higher value on the machine than on the individual, these goals have not been achieved. Instead, technology has been used by the owners of the means of production as an instrument for making ever more profits. This has led to more inequality between rich and poor and destruction of the middle class. If this type of industrial organization is allowed to go on uncontrolled, it will tend to assume autonomy of operation, with further degradation of the human condition.

The emphasis of our society on the material aspects of life is a consequence of the quantitative basis of present science and its branches. Since it is the economic system that primarily interrelates the subsystems of society—it is its source of energy—special attention should be given to the foundations of economic science.

Economic science is a theory of human action, but its statements and propositions are not derived from experience, and its procedures do not follow those of logic and mathematics. It involves judgments, but they are judgments of relevance and not of human values. Since unitary theory was not readily available at the time of its foundation, economic theory uses concepts such as that of "static equilibrium of perfect competition," derived from Newtonian mechanics, and "struggle for survival" ideas derived from Darwinian evolutionary theory. Both of these concepts, however, do not apply directly to human personalities or to socially organized economic systems.

In describing his theory, the Austrian School economist Ludwig von Mises uses static and dualistic concepts and bases his ideas on the misguided philosophy of Freud and other philosophers. Not knowing how a formative process works, the relations of body to mind and individual to society are not properly understood. He envisions a connection of the external to the internal world, but its nature escapes him. As described by Mises, economics is a subjective science, devoid of any values of order or organization.

Man chooses the means (it is a theory of means), whatever they might be, to attain desired ends, selfish or unselfish. This theory of economics is indifferent to the suffering of others. In this sense, it is an inhumane theory.

According to unitary concepts, human action in the economy is the expression of judgments by the individual (or group of individuals), brought about by the workings of the formative process of the brain. We cannot isolate means from ends, because they are both components of the formative process, which, as we have seen, is intrinsically cooperative. A healthy economy should be based on field principles and on a system of social and cultural values and norms of behavior that lead to maximum human cooperation and benefit.

Present economics sees the individual working in isolation, looking only to maximize his profits; unitary economy sees the individual working in cooperation with other individuals and aiming at achieving maximal gains for all. The saying that "the function of banks is to make money" is therefore incomplete. "For the benefit of all participants in the process" should be added.

The present quantitative method of measurement is clearly unsuitable for the requirements of an exact science of living structure. The dimensional variables it uses—length, time, mass—are limited in scope and cannot suitably represent the one-way transformations of structural processes. What is needed is a new, dimensionless formulation of more efficient design, still to be developed, perhaps based on the determination of angular equalities and inequalities and stability and instability of patterns, as proposed by Whyte. Only in this way can standardization of comparison be achieved and the unitary requirement of completeness satisfied.

Since the principles of organic ordering must be established from observation, a new calculus of structural transformations must then be formulated and an appropriate atomic theory developed. Only then can a comprehensive theory of complex structure be achieved. This tremendous amount of work must be done by the cooperative effort of physicists and mathematicians deeply convinced of the validity of the theory and the potential insight it can provide in the resolution of the most basic problems facing science. From this point onward, physics and biology should work together to unravel the mysteries of organic structure and its disease processes. The path to real progress is very hard, but there is no other way out. The human race is very lucky in having the powerful unitary theory at its disposal, which can always be used as a guide.

The reader may have wondered why unitary theory, which describes the process of nature and is applicable to all structure, is not the official one in physics. The major reason is that this powerful science was born out of wedlock, and by this I mean out of academia. It is a bastard theory and, as such, not accepted by Big Physics, by those who have the power to make decisions. Certainly, the works of Lancelot Whyte and Leo Baranski were read and their contents well understood by many prominent academic physicists and biologists, but acceptance of a new theory involves a complex psychological process of adaptation. Perhaps the new generation of leaders may be more willing to accept change. Meanwhile, the whole of civilization is hanging in the balance.

But there are other obstacles to the acceptance of unitary theory that make the reluctance of physicists understandable, although in no way excusable. Acceptance of the theory would be the equivalent of a revolution in physics, a conversion from one type to another, from reversible to irreversible physics. It would be like removing a giant tree by the roots and replacing it with a young plant of differing species, resistance to disease and drought, that still has a lot to grow before beginning to give fruit.

It is not that existing theories are all wrong and no longer valid. They are just viewed under the brighter light of a more general theory. Newton's physics, for instance, is as valid as ever, if it is applied only to stable systems and the same is true for the principle of conservation of energy. Unitary theory does not throw anything away. On the contrary, it reveres all bodies of knowledge, and this includes religion and philosophy, which are components of the civilization process of humanity. It will extract from these disciplines what is good and of value and use it for its further development.

Modern society cannot continue relying on reified concepts that have been dividing us for so long. We must change to veridical ones, which are the basis of unitary science. So we can think of structural patterns instead of numbers, continuity of process instead of intellectual dualisms, and whole truths instead of partial, incomplete ones. These concepts, previously agreed upon by scientific debate, must then be spread to all peoples by the scientific communities of their respective nations. The whole process may take centuries, if not millennia, to be accomplished.

We have today a diversity of nations in several stages of social and intellectual development, but the destiny of our civilization relies on just a powerful few. To the few powerful and most civilized nations belongs the decision of whether to continue on the present path and face inevitable destruction or change course and face a bright future for as long as our planet can sustain us.

Concluding remarks

Since speech was developed, human societies and cultures were guided by unveridical concepts of permanence. For thousands of generations, constant change was recognized in nature, but its continuity could not be grasped, and its inherent order was taken as disorder. Order was thought to be contained in highly abstract, unveridical concepts, imbued with supernatural powers masquerading as process. It took four hundred years of scientific inquiry and the theoretical integration of the gathered evidence to unravel and scientifically explain the natural process, the mysterious entity that constitutes our universe, from which all forms ultimately derive. Today, we can reasonably say that fundamentally there is no matter, no energy, no god, no soul, and no consciousness in physical or organic nature. What really exists is *process*, which is manifested to us in various ways. Life as an entity is, at the very bottom, a manifestation or expression of process.

Death as an entity does not exist. It is the end of process, the attainment of complete or final symmetry.

6

OUR PLACE IN THE UNIVERSAL PROCESS

The human is a special being. It is the final product of a very long series of transformative processes that took place on the surface of our planet, a component of the solar system, itself a part of the larger Milky Way galactic system, a tiny constituent of the universe. The appearance of the human species was preceded by many other species, first of bacteria, then of inanimate plants, and later of animate animals. Such remarkable transformations of form, from simpler to ever more complex organisms, took place over several billion years of evolution and were possible only because there is a hidden formative process continuously at work in nature. This universal process is common to all organisms, and it is at the root of human unity with the rest of organic nature.

Although there are many animals bigger and stronger than man, none possesses his high level of intelligence. This characteristic, together with high manual dexterity, enabled him to invent various tools and to use them to control all other species and dominate his environment. But his dominant position on the planet carries with it a lot of responsibilities.

Humans are, or can be, the most kind or the most dangerous of all animals. Indeed, they have more capacity to destroy themselves than any other species. They require, therefore, a rational system of thought, which only unitary theory can provide. Deprived of unitary thought for many centuries, humans have struggled to discover their proper place in the universe and the right way to live in it.

In spite of his supremacy in nature, man is just an organism and, as such, is dependent on his environment—he is a part of the wider system of organism-in-environment. The inseparable relation of organisms to a specific environment carries with it several limitations, one of which is his transient life.

In the brief discussion that follows, we intend to clarify some misconceptions that have arisen from not considering the human situation as an extension of the universal process. These misconceptions have far-reaching implications that profoundly affect our lives.

Humans as organisms

Living organisms are highly coordinated wholes made up of very delicate structure concentrated in basic units called cells. The cell interior displays a high level of order, but this order is not maximal. There is some room for variation—of fluidity, flexibility, mobility, and so on—which is brought about by stimulation from its exterior and interior environment. With too-weak stimulation, below a given threshold, no change occurs; but with too-strong stimulation, above an adaptive range, living structure is destroyed or not formed in the first place. Organisms are developed and shaped by their respective environments.

The environment-bound nature of metabolic life does not allow organisms to be moved from one environment to another one with differing characteristics. For us to fly, for instance, the ambient pressure and oxygen concentration of the airplane's internal environment must be kept similar to those we are accustomed to on the surface of the earth. With so many environmental variables involved in the living process, such as atmospheric temperature, pressure, and gas concentration, amount and state of water, periodicities, etc., the possibility of humans to be moved, in a distant future, from the earth to an earthlike planet accessible to us is extremely remote. The infinite vastness of the universe, limits imposed on attainable velocity, and the relatively brief human life span are all insurmountable barriers to the exploration of planets possessing physical characteristics similar to those of earth. To endure the passage of time, metabolic life must develop de novo in a suitable environment. It is a bottom-up process that takes billions of years to be accomplished. As organisms, humans can aspire to be no more than what they really are—earth-evolved and earthbound creatures.

The intimate environmental dependency of organisms puts emphasis on the protection of our planet's environment, if we are to survive on it. Moreover, given the relation of everything to everything else, if only one of the environmental variables pertinent to life were to change its magnitude from present levels, it would affect other variables as well. This is the case of the present increase in the atmospheric concentration of carbon dioxide (CO_2), brought about by the oxidation of fossil fuels. In the low layers of the atmosphere, it directly leads to an increased pressure of the dissolved portion of carbon dioxide in our blood, which behaves like an acid; in the high layers of the atmosphere, by preventing the dissipation of solar heat into outer space, it indirectly leads to global warming. Both of these effects, increased acidity of body fluids and increased environmental temperature, are threats to our health and ultimately to our survival.

Another aspect of the nature of life is its continuity. Life is not something that stops and starts again. In special situations, however, the continuity of the field process may temporarily be interrupted. This is the case with germ cells, so it is not correct to assert that life starts at conception. The ovum is very much alive before and after fertilization. Following fertilization, the field latent in the germ cell membrane is activated to start the development of new structure. It is development of the embryo, not life, that starts at conception.

Humans as thinking organisms

Although the human species possesses a thinking organ capable of highly differentiated thought, individual humans are born without knowledge. In this regard, a newly born human brain is a tabula rasa. Each of us must acquire knowledge the best way we can.

Deprived of a proper system for rational thought, our primitive ancestors were guided first by instinct and then, in the course of intellectual development, by a few static and irrational ideas of universal scope. Always fearful for their survival, they became aggressive and cruel. And the vicious circle of fear and aggression never left them, even with the advent of objective scientific thinking. Today, civilized humans continued to be afraid of hunger, war, loss of job, disease, death, etc. We have accepted perpetual fear as a way of living, as something inherent in us that cannot be eradicated.

But with the advent of unitary thinking and the notion of a formative process working toward order inside us, we have reached a stage where we can understand ourselves better than ever before. We know now how our innermost thoughts are brought about, how our feelings arise, and how we can control ourselves at all times. In the unitary system of thought, there is no place for fear or evil. These concepts were formed out of ignorance and can ultimately be conquered by unitary knowledge.

The traditional view that consciousness has a supreme role as controller of behavior is no longer tenable. In effect, this assertion is doubly wrong. First of all, consciousness, as previously described, is not an entity in itself. The valuable element in consciousness is transient attention between organism and a particular stimulus, external or internal. Secondly, our behavior is controlled by an organic tendency and not by any special cause or material agency. It is the dominant formative processes of the brain, themselves expressions of the universal process, that control the general form of behavior. The content of these processes (the particular types of structural aggregates of their memory structures) is influenced by our social environment (our culture), by way of our past experiences. But being predominantly cognitive personalities, humans are capable of ultimately controlling, through their own will, their own behavior, except in extreme circumstances when unusually strong emotions may override the intellect.

Since humans are a symbolic class of life, all human behavior is conditioned, shaped, and mediated by symbols. The proper symbols should be general ideas, but when they are not available, the individual or the community resorts to the symbols of the father, the group leader, the hero, or the role model. But a full understanding of the universal form of process can enable humans to overcome personal frustration and find their own roles in society.

Humans as social beings

In unitary science, a given system is always a component of a wider system. In the case of the human system, the wider system is its surrounding environment, physical and social. The social environment is made up of an extreme variability of individuals, all at different stages of social, economic, and cultural development. This is also the case of nations. But in spite of these differences, all humans have something in common—they all belong to the human race.

And the race, for the benefit of all members, should be united. But several attempts at unification, by religions and various philosophical doctrines, all have failed. Deep down, human societies and nations continue to be divided. Perhaps the unveridical concepts promoted by the above educational systems are too rudimentary or incomplete to be accepted by all people. Whatever the reason, under these circumstances, cooperation in the resolution of extremely important issues to mankind—energy crisis, climate change, atomic and biological weapons, water pollution and commercialization, food production, population growth, biodiversity loss, and human rights, to name just a few—is very difficult or impossible to achieve, and mutual development is not facilitated.

But why not formulate a universal standard of values acceptable to everyone? The answer to this question is that although it is desirable, this is impossible to accomplish. There is no supreme authority, recognized by all humankind, capable of formulating it, and even if there were, no values can be appropriate to all, since individual and communities vary greatly. What is acceptable to some may not be so to others.

What unites people are unifying ideas, powerful enough to withstand critical scrutiny and flexible enough to be modified if necessary. These ideas must first be accepted by science without reservations, and for this to take place, they must express the scientific truth. An idea of such power is that contained in the unitary theory here described, which expresses, as far as it is known, the law of nature.

To prevent our civilization from destruction, it is of practical necessity that, in the near future, the nations of the world be united under the conviction of a formative process in nature that includes all humanity. Only in this way can the above social problems, and many others, be resolved in time to avoid destruction of the human race by the race itself.

In the long run, with proper education of all individual humans and reorganization of the present tradition, we may be capable of finally developing a world free of fear, war, and unemployment, and a civilization emphasizing scientific reasoning, human values, and mutual development.

Humans as transient forms of life

The universal field process goes on eternally, but any form separated from the field must sooner or later regress to it. All forms are transient, including human forms. We continue in process, but it is a sideline process of limited duration. It will ultimately come to an end, and when that happens, death becomes apparent.

Death is not a punishment for anything we have done wrongly in life; it is a natural event, the final one for us. As persons, we are not only members of a family and citizens of a state; we are, above all, a part of nature and, as such, must follow its laws. We came from nature, and we must go back to nature—this is the cycle of life.

When the meaning of the life cycle is fully understood and we unreservedly accept the inevitability of death, we become at ease with ourselves—there is no longer fear. We free ourselves from the heavy burden of the thought of death, and only then can we live an authentic, joyous life.

But is there any purpose in living? The answer to this question is that there is no specific purpose in life. Life, however, is a process, and as such, we should facilitate its development; that is, the development of ourselves and the development of others.

On a deeper thought, it is the finiteness of life that stimulates us to make choices, to accomplish what is possible, and to distinguish ourselves. If we were immortal beings, there would be no need to make one choice over another, for we would have an infinite amount of time to experience all possibilities. One day would be no more or less significant than the next. Without death, life would lose all its dignity.

"Man should not die without having lived, without having experienced enough joy to bring him peace when the end is near," wrote Whyte. Almost a score of centuries earlier, Seneca (4 BC to AD 65), a Stoic Roman philosopher, had similar thoughts on living: "The greatest obstacle to living is expectancy, which hangs upon tomorrow and loses today. You are arranging what lies in Fortune's control and abandoning what lies in yours. What are you looking at? To what goal are you straining? The whole future lies in uncertainty: live now." Both views represent spontaneous expressions of the formative process of nature at work inside all of us.

SELECTED REFERENCES

Chapter 1

1. Baranski, L. J. *Scientific Basis for World Civilization: Unitary Field Theory*. Boston: The Christopher Publishing House, 1960.
2. Carvalho, J. S. *From Stars to Societies: A Process in Time*. Baltimore: American Literary Press, 2009.
3. Grundbaum, A. "A New Critique of Theoretical Interpretations of Physical Cosmology." *British Journal for the Philosophy of Science* 51 (2000): 1–43.
4. Whyte, L. L., ed. *Aspects of Form: A Symposium on Form in Nature and Art*. New York: American Elsevier Publishing Company, Inc., 1968.
5. Whyte, L. L. *The Unitary Principle in Physics and Biology*. London: The Cresset Press, 1949.
6. Whyte, L. L. *The Next Development in Man*. New York: Mentor Books, 1962.

Chapter 2

1. Baranski, L. J. *Scientific Basis for World Civilization: Unitary Field Theory*. Boston: The Christopher Publishing House, 1960.
2. Bowler, P. J. "Darwin's Originality." *Science* 323 (2009): 223–226.
3. Bulkley, D. H. "An Electromagnetic Theory of Life." *Medical Hypotheses* 30 (1989): 281–285.
4. Hollister, P. *Origin and Evolution of the Universe: A Unified Scientific Theory*. http://www.origin-of-universe.com.

5. Jacobson, J. I. *The Secret of Life: Perspectivism in Science.* New York: Philosophical Library, Inc., 1983.
6. Nussinov, M. D. and V. I. Maron. "Impulse Paradigm of Self-Organization of Matter in the Universe, and Nanobiological Principles." *Nanobiology* 2 (1993): 215–228.
7. Watterson, J. G. "The Interactions of Water and Proteins in Cellular Function." *Progress in Molecular and Subcellular Biology* 12 (1991): 113–134.
8. Whyte, L. L. *Archimedes, or the Future of Physics.* New York: E.P. Dutton & Company, 1928.
9. Whyte, L. L. *The Unitary Principle in Physics and Biology.* London: The Cresset Press, 1949.
10. Whyte, L. L. *Action on Form.* Westport, Connecticut: Greenwood Press, 1954.
11. Whyte, L. L. *The Universe of Experience.* New York: Harper & Row, 1974.
12. Wiggins, P. M. "Enzyme Reactions and Two-State Water." *Journal of Biology and Physical Chemistry* 2 (2002): 25–37.
13. Wiggins, P. M. "The Source of Some of the Extraordinary Powers and Properties of Enzymes." *Water* 1 (2009): 35–41.

Chapter 3

1. Baranski, L. J. *Scientific Basis for World Civilization: Unitary Field Theory.* Boston: The Christopher Publishing House, 1960.
2. Hameroff, S. R. and R. Penrose. "Conscious Events as Orchestrated Space-Time Selections." In *Explaining Consciousness: The "Hard Problem,"* edited by Jonathan Shearer, 177–195. Cambridge, Massachusetts: MIT Press, 1997.
3. Mahowald, M. W. "Understanding Consciousness." *Perspectives in Biology and Medicine* 46 (2003): 297–304.
4. Nichols, S. "Experimental Philosophy and the Problem of Free Will." *Science* 331 (2011): 1401–1403.

5. Shear, Jonathan, ed. *Explaining Consciousness: The "Hard Problem."* Cambridge, Massachusetts: MIT Press, 1997.
6. Sieb, R. A. "The Emergence of Consciousness." *Medical Hypotheses* 63 (2004): 900–904.
7. Tenenbaum, J. B., C. Kemp, T. L. Griffiths, and N. D. Goodman. "Proprioception: Is the Sensory System That Supports Body Posture and Movement Also the Root of Our Understanding of Physical Laws?" *Nature* 432 (2004): 21.
8. Tenenbaum, J. B., C. Kemp, T. L. Griffiths, and N. D. Goodman. "How to Grow a Mind: Statistics, Structure, and Abstraction." *Science* 331 (2011): 1279–1285.
9. Whyte, L. L. *Time and the Mind-Body Problem: A Changed Scientific Conception of Process.* Zurich: Rhein-Verlag, 1952.
10. Whyte, L. L. *Focus and Diversions.* New York: George Braziller, Inc., 1963.
11. Whyte, L. L. *The Unconscious Before Freud.* New York: St. Martin's Press, 1978.

Chapter 4

1. Ambrose, S. E. *Undaunted Courage.* Pleasantville, New York: Reader's Digest Association, 1996.
2. Baranski, L. J. *Scientific Basis for World Civilization. Unitary Field Theory.* Boston: The Christopher Publishing House, 1960.
3. Hayakawa, S. I. *Language in Action. New York: Harcourt, Brace & Company, Inc., 1939.*
4. Hayakawa, S. I. *Symbol, Status, and Personality.* New York: Harcourt, Brace & World, Inc., 1963.
5. Premack, D. "Is Language the Key to Human Intelligence?" *Science* 303 (2004): 318–320.
6. Whyte, L. L. *Accent on Form.* Westport, Connecticut: Greenwood Press, 1954.
7. Whyte, L. L. *The Next Development in Man.* New York: Mentor Books, 1962.

8. Wilczek, F. "When Words Fail." *Nature* 410 (2001): 149.

Chapter 5

1. Charlton, B. and P. Andras. *The Modernization Imperative: A Systems Theory Account of Liberal Democratic Society.* Exeter, UK: Imprint Academic, 2003.
2. Darwin, C. and A. R. Wallace. "On the Tendency of Species to Form Varieties, and on the Preservation of Varieties and Species by Natural Selection." *Journal of the Proceedings of the Linnean Society.* Zoology 3 (1858): 46–50.
3. Goldin, C. and L. F. Katz. *The Race between Education and Technology.* Cambridge, Massachusetts: Belknap Press, 2008.
4. Hayakawa, S. I. *Language in Thought and Action.* New York: Harcourt, Brace and Company, 1949.
5. Lawler, A. "Uncovering Civilization's Roots. *Science* 335 (2012): 790–793.
6. Mises, L. V. *Human Action: A Treatise on Economics.* San Francisco: Fox and Wilkes, 1996.
7. Soros, G. *The Age of Fallibility.* Cambridge, Massachusetts: Public Affairs, 2006.
8. Weintraub, E. R. "First, Kill the Economists." *Science* 320 (2008): 318–319.
9. Whyte, L. L. Every Man Looks Forward. Henry Holt and Company, New York, 1948.
10. Whyte, L. L. *The Atomic Problem: A Challenge to Physicists and Mathematicians.* London: George Allen and Unwin Ltd., 1961.

Chapter 6

1. Annas, G. J. *Worst Case Bioethics: Death, Disaster, and Public Health.* New York: Oxford University Press, 2010.

2. Campbell, R. and C. D. N. Costa. *Seneca: Letters from a Stoic and Three Dialogues.* Penguin Books Ltd., 2003.
3. Earnshaw, S. *Existentialism: A Guide for the Perplexed.* New York: MJ Books, 2006.
4. Krauss, L. M. "Space, the Final Frontier?" *Scientific American* 301 (2009): 36.
5. Pelosfsky, S. and R. Pelofsky. "The Voice of Art and the Art of Medicine." *Journal of Neurosurgery* 97 (2002): 1261–1268.
6. Whyte, L. L. *The Next Development in Man.* New York: Mentor Books, 1962.
7. Whyte, L. L. *Internal Factors in Evolution.* New York: George Braziller, 1965.
8. Whyte, L. L. *The Universe of Experience.* Harper & Row, 1974.

INDEX

Made in the USA
Monee, IL
28 February 2026

45072220R00085